P9-EKS-682

UNIT PROCESSES SERIES
ORGANIC CHEMICAL INDUSTRIES
VOLUME 1

UNIT PROCESS GUIDE TO ORGANIC CHEMICAL INDUSTRIES

UNIT PROCESSES SERIES
ORGANIC CHEMICAL INDUSTRIES
VOLUME 1

UNIT PROCESS GUIDE TO ORGANIC CHEMICAL INDUSTRIES

by

ELBERT C. HERRICK
JOHN A. KING
ROBERT P. OUELLETTE

MITRE Corporation
McLean, Virginia

and

PAUL N. CHEREMISINOFF

New Jersey Institute of Technology
Newark, New Jersey

ANN ARBOR SCIENCE
PUBLISHERS INC
P.O. BOX 1425 • ANN ARBOR, MICH. 48106

P.O. Box 1425, Ann Arbor, Michigan 48106

Library of Congress Catalog Card No. 79-89444
ISBN 0-250-40328-5

Manufactured in the United States of America

PREFACE

The manufacturing processes used in the organic chemical industry consist of "unit process" components that carry out the fundamental chemical reactions in organic synthesis. These unit processes are the basic building blocks of organic chemical manufacturing. This *Unit Process Guide* identifies the 39 commercially significant unit processes used in the manufacture of the 263 principal organic chemicals in U.S. commerce.

This *Unit Process Guide* was prepared by: (1) selecting 263 commercial organic chemicals synthetically produced in the U.S.; (2) identifying each of the unit processes used in the manufacturing routes for each of the 263 selected organic chemicals; and (3) identifying the processes available for license to manufacture these 263 chemicals. This volume verifies that 39 unit processes in commercial use may be meaningfully defined.

The 39 unit processes have been grouped into 23 major unit processes used in large-volume commercial production and 16 minor unit processes used in smaller-scale production. Tables for each unit process list the products manufactured using this process, the feedstocks required, other unit processes involved in the manufacture of the product, the number of U.S. commercial plants making the product, the owners of processes available for license, and the number of worldwide plants using each licensed process. One table lists the 263 commercial organic chemicals alphabetically; it includes the feedstocks required and the unit processes involved in the manufacture of each chemical.

The advantage of the approach used in this book is clear: it reaffirms the unit process method of classifying toxic impurities in chemical feedstocks, intermediates and products as they are moved in commerce. It is felt that engineers will be more motivated to accept this mid-course correction than the classical industry-by-industry, or priority pollutant, approach employed to date. It is hoped that the chemical manufacturing industry will appreciate this technique and reap its benefits.

The authors gratefully acknowledge the U.S. Environmental Protection Agency, whose support under EPA Grant R805630-01-1 in part made this work possible. Specifically, our thanks to David R. Watkins and Paul E. desRosiers of the U.S. Environmental Protection Agency, for their encouragement and assistance.

Elbert C. Herrick
John A. King
Robert P. Ouellette
Paul N. Cheremisinoff

AUTHORS

ELBERT C. HERRICK is a technical staff member of the METREK Division of the MITRE Corporation and has been intimately involved with the identification of toxic pollutant discharges for unit processes. A graduate of Montana State University and Princeton University, he received his PhD in organic chemistry from Massachusetts Institute of Technology in 1949. He has had wide industrial experience with such companies as Sun Oil, Dow Chemical Company and as Director of Chemical Research for Escambia Chemical Corporation. He is a member of a large number of technical and honor societies, including Sigma Xi, Tau Beta Pi and Phi Kappa Phi.

JOHN A. KING is a technical staff member of the METREK Division of the MITRE Corporation. Prior to joining MITRE, Mr. King was associated with Tracor/Jitco, McGraw Hill Publications, Vitro Labs and J. T. Baker Chemical Company. He is a graduate of Polytechnic Institute of Brooklyn and has had more than 20 years of hands-on oil refinery/chemical manufacturing experience in industrial management, as well as technical data analysis. Widely published in the technical press, he was editor of the Ann Arbor Science Publishers' book *Electrotechnology Vol. 1–Wastewater Treatment and Separation Methods.*

ROBERT P. OUELLETTE is Associate Technical Director, Energy, Resources and the Environment, for METREK, a Division of the MITRE Corporation. Dr. Ouellette has been associated with the MITRE Corporation in varying capacities since 1969, and has been Associate Technical Director since 1974. Prior to joining MITRE he was with TRW Systems, Hazelton Labs, Inc. and Massachusetts General Hospital. A graduate of the University of Montreal, he holds a PhD from the University of Ottawa, and has published a considerable number of technical papers and books in the energy-environment areas. His memberships include the American Statistical Association, Biometrics Society, Atomic Industrial Forum and the NSF Technical Advisory Panel on Hazardous Substances.

PAUL N. CHEREMISINOFF is Associate Professor of Environmental Engienering at the New Jersey Institute of Technology. A consultant and registered professional engineer, he has more than thirty years of practical design, development and manufacturing engineering experience in a wide range of organizations, specifically in chemical processing. He is the author/editor of many Ann Arbor Science Publishers handbooks, including *Pollution Engineering Practice Handbook, Carbon Adsorption Handbook* and *The Environmental Impact Data Book.* Also, he is Engineering Editor of *Water & Sewage Works,* an international journal.

TABLE OF CONTENTS

LIST OF FIGURES

LIST OF TABLES

1

UNIT PROCESSES AND ENVIRONMENTAL REGULATIONS

INTRODUCTION

More than 10,000 chemical manufacturing and processing firms employ 1.1 million Americans, with sales of about $100 billion. Chemical industry products represent some 7.5% of the U.S. Gross National Product (GNP). Over the past 10 years, the industry has enjoyed an annual growth rate nearly twice that of the GNP. New chemicals represent about 10% of all patents issued in the United States from 1836 to 1970. Today, there are more than 30,000 chemical substances in commerce and another 3-4 million in research and development; as many as 1,500 new chemicals enter the market every year.

As a result of environmental regulations enacted in recent times, both the toxic pollutants and chemical products are now regulated, forcing the U.S. Environmental Protection Agency (EPA) to consider the intraprocess sources of toxic chemical discharges to the air, water and solid media, as well as the intraprocess sources of toxic contaminants. There is an enormous multiplicity of chemical manufacturing operations, and the task associated with environmental assessment of so many processes is awesome. Therefore, EPA may consider assessment of the generic "unit process" reactions–the building blocks for chemical manufacturing operations for regulatory purposes.

This *Unit Process Guide* identifies the 39 commercially significant unit processes used in more than 5,000 plants to manufacture 263 principal chemiclas in U.S. commerce. It provides the basis to begin systematic environmental assessment of the 39 unit processes as they are utilized in the chemical industry today.

Forthcoming volumes will evaluate the pollutant and product contamination risks of the 39 unit processes to facilitate environmental assessment for the 263 primary organic chemicals of commerce.

REGULATORY BACKGROUND

Today, 12 federal statutes control pollution and human and environmental exposure to toxic chemicals (Table 1.1).

Table 1.1 Federal Laws Regulating Toxic Chemicals

Title	Abbreviation	Public Law No.
Toxic Substance Control Act of 1976	TOSCA	94-469
Food, Drug and Cosmetic Act as amended in 1976	FDCA	94-295
Occupational Safety and Health Act of 1970	OSHA	91-596
Consumer Product Safety Act of 1970	CPSA	92-573
Marine Protection, Research and Sanctuaries Act of 1972	Ocean Dumping	92-532
Federal Pesticide Act of 1978	FPA	95-396
Clean Air Act as Amended in 1977	CAA	95-95
Federal Water Pollution Control Act,	FWPCA	92-500
amended as Clean Water Act of 1977	CWA	95-217
Safe Water Drinking Act of 1974	SDWA	93-523
Resource Conservation and Recovery Act of 1976	RCRA	94-580
Hazardous Materials Transportation Act of 1970	HMTA	91-458
National Environmental Policy Act of 1969	NEPA	91-190

Regulatory action required by these statutes focuses on toxic chemicals in wastes and products, their ambient and occupational environments, identification, sources, control, safe handling, discharge to the environment and ultimate disposal. A hypothetical regulatory scenario of the future, assuming regulations will be issued under all 12 statutes, is depicted in Figure 1.1.

As shown in Figure 1.1, the projected modes of regulation for toxic chemicals in the future are exceedingly complex:

1. Consumer uses of chemicals may be regulated under the Consumer Product Safety Act; Toxic Substances Control Act; Federal Pesticide Act; and Food, Drug and Cosmetic Act, depending on the type of product or mode of use.
2. Workplace conditions in power and manufacturing plants are regulated by the Occupational Safety and Health Act; water discharges by the Clean Water Act; solid discharges by the Resources Conservation and Recovery Act; and air discharges by the Clean Air Act.
3. Air pollution control facilities emissions are regulated under the Clean Air Act; product solid wastes are regulated under the Resources Conservation and Recovery Act; and wastewaters (scrubber waters) are regulated by the Clean Water Act.

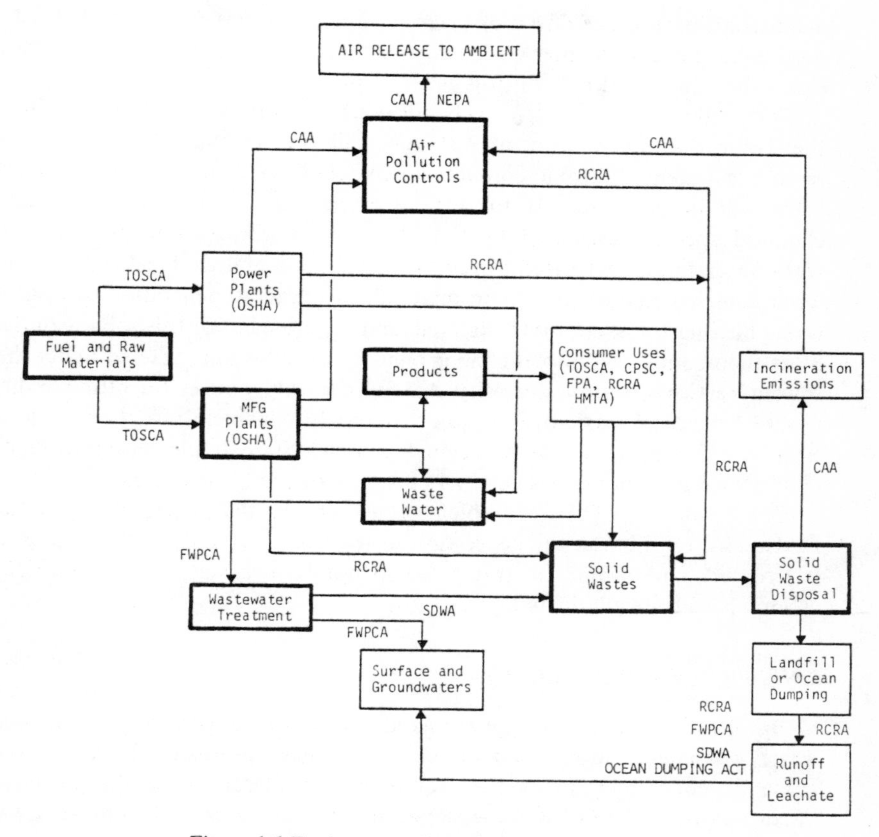

Figure 1.1 Environmental regulation scenario.

4. Water pollution control facilities, both industrial and municipal, discharge treated waters to navigable waters under Clean Water Act regulations and to drinking water supplies under the Safe Drinking Water Act. Water pollution control facilities generally generate solid wastes regulated under the Resources Conservation and Recovery Act.
5. Solid waste disposal procedures may be regulated by:
   (a) the Ocean Dumping Act,
   (b) the Clean Air Act (air emissions from incineration), or
   (c) the Resources Conservation and Recovery Act and Safe Drinking Water Act (due to ultimate potentials for toxic chemical impacts on drinking water supplies).

There are two principal problems created by this scenario: (1) toxic materials, potentially accumulate at various points in the chemical manufacturing use and disposal cycle in concentrations or forms that may pose toxic hazards to mankind; and (2) the regulations may conflict with each other, *i.e.*, pollution controls generate pollution.

For example, in the case of a chemical manufacturing operation for which the Toxic Substances Control Act may call for washing of a product to remove a potentially toxic contaminant, wastewater containing the contaminant will be generated. If the toxic material falls within one of the 131 chemical species designated by the Clean Water Act, Section 307 (a), the wastewater must be treated to remove the toxic material. Treatment of the wastewater to remove the toxic material may lead to generation of a solid waste bearing the same toxic material, and management of this solid material must then comply with forthcoming regulations to be administered under the Resources Conservation Recovery Act. If the only option for ultimate disposal of the solid waste bearing toxic materials is incineration, the emissions from the combustion will be regulated under State Implementation Plans (SIP) issued in response to Section 111 or 112 of the Clean Air Act.

One resolution of these problems is to modify the process itself, or the feedstocks, to eliminte the generation of the toxic materials. If this approach is adopted, end-of-pipe treatment would not be necessary, since toxic discharges have been eliminated.

UNIT PROCESS REGULATION

By virtue of proposed regulations affecting the feedstock, products and waste discharges, EPA could effectively regulate processes. EPA received guidance concerning process or equipment standards under the amended Clean Air Act of 1977. Congressional intent concerning such standards was expressed as follows:

> "The Committee recognizes that in some instances [such as control of emissions from petroleum liquid storage vessels, or from organic compounds of carcinogenic potential] it may not be possible for the Administrator to promulgate and enforce a performance standard. While the Committee intends that a performance standard be promulgated whenever practicable, the Committee does not intend to prevent the Administrator from dealing with problems where this would be impracticable . . . Moreover, the Committee expects the Administrator to include numerical performance standards whenever technological advances, improved measurement methods or other changed circumstances made numerical standards practicable."

Under the Clean Air Act, a performance standard is a "standard for emissions of air pollutants which reflects the degree of emission limitations achievable through the application of the best system of emission reduction."

The intent of the statute is that EPA should determine the achievable limits and let the manufacturer select the process. In one instance, EPA did issue a process standard, which applies to storage vessels for petroleum liquids. EPA indicated that only an equipment or process standard can be applied when the source does not lend itself to emissions testing.

The Toxic Substances Control Act (Section 6a) requires EPA to protect the public against unreasonable risk of injury to health or to the environment posed by a chemical substance or mixture in a product. Section 6a could also lead to process standards or guidelines. The criteria used to define unreasonable risk will help determine how this section will be used in regulating chemicals and processes. For example, if unreasonable risk means a substantial increase in mortality in the general population, this law will be less widely applicable than if unreasonable risk is defined as impairment of health in a susceptible part of the population.

Section 6b of the TSCA states that the Agency may order revisions in quality control procedure, if adulteration of the chemical substance is occurring due to inadequate process control. Stipulated improvements in quality control may involve in-process or equipment changes. However, regulatory actions under this section may be taken only for particular manufacturers or processes, not manufacturers in general. Moreover, this law may be used only to prevent unreasonable risk associated with adulteration of products, not risk associated with pollutant discharges. Therefore, any use of 6b to impose pollution control must be for the expressed intent of product decontamination controls.

Under Section 307 (a) of the 1977 Clean Water Act, rule-making activities to specify pollution control for some 131 water pollutants could conceivably involve specification of in-process modifications. However, recent delays of regulatory activity preclude such a toxics regulatory approach.

While regulatory authority to stipulate use of a given manufacturing process may not directly exist under any of these laws, a manufacturer may not have many processing options in a situation in which a toxic chemical exists in a feedstock, waste or product. Regulation of all inputs and outputs in a process implies process regulation. Therefore, EPA faces many issues associated with regulating toxic chemicals and needs to identify those toxic chemicals of serious concern that may occur in feedstocks, products and waste discharges, as the result of the basic process configuration.

There are thousands of processes in thousands of manufacturing operations used in industry; these involve many thousands of chemical products and effluents. Evaluation of all these processes for each of the many potentially toxic materials on an operation-by-operation basis poses practically an impossible task. However, for the organic chemical industries, there are some 39 unit processes comprising the bulk of manufacturing operations.

If the unit processes used in a given industry segment or plant are identified, and if typical toxic discharges and product contaminants for these unit processes are indicated, there would be an opportunity to focus regulatory activities on 39 unit processes instead of thousands of manufacturing processes.

The primary objective of this book is to identify which unit processes are in use and to indicate likely toxic pollutant discharges and product contaminants for each of these unit processes. The first step is identification of the unit processes in commercial use.

The next step will involve evaluating each unit process from the standpoint of toxic chemical occurrence in feedstocks, products and waste discharges. The approach will involve selection of representative manufacturing operations for each of the unit processes. Analysis of results for each unit process will lead to identification of common toxic discharges and product contamination or their absence. Following this evaluation for air, water and solid discharges and for product contamination, plant site visits and meetings with industry could be used to verify the results and to facilitate a survey of available discharge-treatment techniques. The charcterization of toxic materials in discharges and wastes, along with the survey of available treatment techniques, will then lead to definition of needs for development of new treatment techniques and process modifications.

The end products of the total effort will be:

1. evaluation/identification of toxic discharges and product contaminants for each of 39 unit processes; and
2. indication of research and development needs for each unit process.

The first of these two primary products can be used for guidelines and standards development, as well as for subsequent enforcement of these regulations by EPA's regional offices. The data will serve as a baseline for process pollution control needs identification and contribute to focused enforcement of the amended Clean Water Act, the Toxic Substances Control Act, Resources Conservation and Recovery Act and Clean Air Act provisions directed at specific toxic materials.

For example, for each unit process we will identify chemicals likely to occur in wastewaters, solid discharges, air emission and, in some cases, as contaminants in products. Once the chemicals are identified, the toxic risk can be assessed for the modes of human and environmental exposures that could occur. If toxic risks are considered significant, methods of risk abatement must be considered. These risk abatement methods will include:

- pollution control technologies
- product modification

- feedstock quality control or switching
- changes in process conditions

For each of these methods, the research and development needs prerequisite to their implementation may be identified. Furthermore, through this assessment, the media for which significant toxic risks commonly occur may be pinpointed for consideration in implementing the air, water, solid waste and product regulations.

RATIONALE

Most of the manufacturing processes widely used in chemical industry today consist of "unit process" components that carry out the fundamental chemical reactions in organic synthesis, *e.g.*, nitration, amination by ammonolysis, etc. Unit processes are the basic building blocks of chemical manufacturing operation. For most commercial applications of a given unit process, the physical and organic chemistry tends to be common or alike.

Through systematic examination of the process chemistry associated with the unit processes, one can identify the air, water and solids discharges, as well as the potentials for product contamination for commercial chemical manufacturing operations.

The first challenge in carrying out such an assessment was to survey the commercially used chemical manufacturing operations to determine: (1) which unit processes are in commercial use, and (2) which unit processes are used for each commercial application. During the survey, it also became apparent that these unit processes, fall into two major groups: (1) major unit processes used in large-volume chemical production, and (2) minor unit processes used in smaller-scale, yet significant, production. These two groups of unit processes are listed in Tables 3.1 and 3.3.

Because of the commonality of physical and organic chemistry and the discharges associated with the use of the unit processes, they should be the focus of an assessment of potentials and risks for toxic waste discharges and product contamination in the chemical industry. Many manufacturing operations used throughout the process industries, *i.e.*, chemical, textile, paper, rubber, plastics, dyes and detergents may be environmentally assessed through unit process assessment.

2

METHODS USED TO CREATE THE GUIDE

INTRODUCTION

There were three steps involved in creating the *Unit Process Guide:* identification of (1) the 263 selected commercial chemicals synthetically produced in the U.S.; (2) the unit processes used in the manufacturing routes for each of the 263 selected chemicals; and (3) the commercially owned and used manufacturing processes.

Many manufacturing synthesis routes involve more than one unit process. For example, the manufacture of toluene diisocyanate, starting with toluene, involves three unit processes:

- *nitration* of toluene to dinitrotoluene
- *hydrogenation* of dinitrotoluene to diaminotoluene
- *phosgenation* of diaminotoluene to toluene diisocyanate

There are two reported commercial toluene "dinitration" processes—one owned by Meissner and the other owned by Sumitomo. The dinitrotoluene "hydrogenation" and the diaminotoluene "phosgenation" processes are owned by Allied, FMC, Nippon Soda-Nissan and Sumitomo.

This example illustrates that there may be multiple unit processes for a given manufacturing synthesis route and multiple commercially owned and used manfuacturing processes for each unit process. Accordingly, there are three entities involved in this catalog:

1. *Manufacturing synthesis routes:* a list of the unit processes used in the synthesis route, *e.g.*, nitration, hydrogenation and phosgenation unit processes used in the manufacturing synthesis route from toluene to toluene diisocyanate;
2. *Unit Processes:* discrete, identifiable, individual chemical reactions used commercially in organic synthesis, *e.g.*, nitration, hydrogenation and phosgenation; and

3. *Manufacturing Processes:* a list of the commercially owned and/or licensed manufacturing processes used to carry out the application of a unit process, *e.g.*, Sumitomo and Meissner for toluene nitration.

IDENTIFICATION OF THE PRINCIPAL COMMERCIAL CHEMICALS

The 263 commercially significant chemicals were selected using the following:

1. list of 131 chemicals on the EPA Recommended List of Priority Pollutants, pursuant to the Consent Decree of December 5, 1977 and the Clean Water Act of 1977, Section 307(a);
2. list of 88 chemicals surveyed and ranked for 1982 Projected Total Annual Gross Emissions from the Hydroscience, Inc. Progress Report No. 7 for September 1-September 30, 1977 on "Emissions Control Options for the Synthetic Organic Chemicals Manufacturing Industry," under EPA Contract No. 68-02-2577;
3. list of 59 organic chemicals being studied by Environmental Science and Engineering, Inc., under an EPA contract aimed at evaluating carbon adsorption pollution control;
4. classification of 16 nitration-products/processes, from "Air Pollution from Nitration Processes" by Processes Research, Inc., under EPA contract No. CPA 70-1, March 31, 1972.
5. classification of 25 chlorination/hydrochlorination processes, from "Air Pollution from Chlorination Processes," by Processes Research, Inc., under EPA contract No. CPA 70-1, March 31, 1972.
6. products and by-products of chemical process technology available for license or sale, from "Sources and Production Economics of Chemical Products," *Chemical Engineering,* McGraw-Hill Publications Co., 1974.

These lists were combined by placing each chemical in a file with all primary synonyms. Those chemicals and sources not included in the 1977 SRI-Directory of U.S. Chemical Producers were eliminated from the file as noncommercial-scale chemicals.

IDENTIFICATION OF UNIT PROCESSES

The manufacturing synthesis routes were identified for each of 263 selected commercially produced chemicals through telephone survey, search of the literature and manufacturing companies.

The main chemical reactions (nitration, hydrogenation and phosgenation) in each manufacturing synthesis route were identified. Some 23 unit processes were identified as widely used for large-volume manufacturing operations and were designated as major unit processes. Likewise, some 16 minor processes (used in small-volume manufacture) were identified and used to characterize manufacturing synthesis routes. In some instances, rather rigorous definition of unit processes were necessary to avoid undesirable increase in the number of unit processes and multiple entry under ambiguously defined unit processes. Chapters 4 and 5 provide the defintiions used as the bases for identifying and listing the unit processes that were used to characterize each manufacturing synthesis route.

IDENTIFICATION OF MANUFACTURING PROCESSES

Simultaneously with the foregoing step, the owners and/or licensors of some 658 manufacturing processes used to commercialize the unit processes were identified using telephone survey and the source references. The 145 owner/licensor companies and their addresses are listed in Chapter 6. Many of the commercially used manufacturing processes are owned by foreign companies.

STRUCTURE AND USES OF THE GUIDE

Major Unit Processes

Table 3.1 lists the 23 major unit processes, the number of compounds produced using these processes (total 410), and entries to the table (total 437) for each process. Table 3.2 lists under each process the chemicals used in their manufacture.

The major processes are used in the manufacture of a number of different chemicals, both in large volumes and in many plants. The only exception is Steam Reforming–Water Gas Reaction, which yields methanol as the only major product. However, as methanol is made by this process in 12 U.S. plants with a capacity of 1.4 billion gallons, it was included among the major processes.

The processes are listed alphabetically in Table 3.2, and the major product is listed alphabetically within each process. The contents will be illustrated using entries from the section dealing with Alkylation. The first entry for acetic acid shows that the oxidation process is involved as well as alkylation. The feedstock is n-butenes, and the owner of a process available for license is Bayer AG. No licensed commercial plants are in operation, as shown by "N" in the fifth column. The last column shows that there are 10 commercial plants operating in the U.S. This number is independent of the columns on "Owner of Process" and "No. of Licensed Commercial Plants." These refer to worldwide licenses, so the numbers may be considerably higher than for U.S. plants.

Linear alkyl benzenes are made from two different feedstocks and are listed accordingly. Phenol and acetone are products from the feedstocks benzene and propylene. The process is indexed under phenol, since acetone is a by-product. Here, four processes are involved–Alkylation, Oxidation, Hydrolysis and Acid Cleavage. The product is cross-indexed under each of these processes. If only one process is required, as in alkylation of benzene with ethylene to ethylbenzene, the word "None" is entered in Column I.

Minor Unit Processes

Table 3.3 lists the 16 minor unit processes, the number of compounds made by these processes (total 38), and entries to the table (total 36) for each process. Table 3.4 follows with the chemicals listed alphabetically under each process used in their manufacture.

Most of these minor unit processes are used in the manufacture of only a few chemicals, which are generally made in relatively low volumes. They may be one step in a sequence, such as Acid Cleavage for phenol-acetone manufacture. Some are very specific, such as Oximation and the Beckmann Rearrangement, which are used in the manufacture of caprolactam from cyclohexanol/cyclohexanone, ammonia and oleum.

One change in the tabular format was required from Table 3.2. The first column is entitled "Processes." The word "None" is not used here because it is not applicable. The name of the minor unit process is repeated as required. For example, Carboxylation is used for salicylic acid and again for sodium *p*-aminosalicylate, because that is the only process involved. Acid Cleavage is not repeated for the second phenol-acetone entry but is involved as one of the four processes.

Index of the Chemicals Produced by Unit Processes

Table 3.5 is an alphabetical listing of all the chemicals from Tables 3.2 and 3.4. There are 263 commercial organic chemicals tabulated in this index.

Entries are made for each feedstock, but the product is only counted once. For example, No. 1, acetaldehyde, is listed twice; No. 2, acetic acid, is listed six times; No. 3, acetic anhydride, is listed twice.

There are 374 product entries by feedstock for the 263 chemicals in the index. For instance, the 3 chemicals listed above are listed 10 times for the 10 processes involving different feedstocks in each of the 3 cases. As shown by Table 3.1 and 3.3, there are 437 entries to the table of major processes and 36 entries to the table for minor processes, for a total of 473 entries. This higher total results from an entry being made to the table for each of the processes used in manufacture of a chemical. For example, phenol/acetone requires four processes and is listed under Alkylation, Oxidation, Hydrolysis and Acid Cleavage. All these processes are cross-indexed for each product in Tables 3.2, 3.4 and 3.5.

How to Use the Unit Processes Guide

The intent of the book is to identify the unit processes used to produce the 263 primary synthetic organic chemicals, so that assessment of the toxic chemical pollutant discharge and product contamination risks for these organic chemical manufacturing operations can focus on their fundamental processes.

Once the unit processes are identified, assessment of pollution discharge and product potentials for each unit process may be applied to the assessment of the overall manufacturing operation's environmental effects. Assessments of pollution and product contamination for the 39 unit processes, now just beginning, will be forthcoming during the next three years.

The *Unit Process Guide* has other uses. Requirements for control technology development will evolve as untreatable waste streams are identified by examining the 39 primary organic chemical synthesis routes. Moreover, in cases in which a manufacturing plant is being planned, a process engineer may use the *Guide* to select a sequence of unit processes that impose minimal pollution discharges. If a process engineer wished to know the state-of-the-art for manufacture of a given chemical, he can find the chemical product in Table 3.5, which lists the unit processes used for each of 263 chemicals. Then he can identify the owners, or licensors, of existing processes, the feedstocks used and, in some cases, the number of plants in the U.S. and abroad by using Tables 3.2 and 3.4.

3

UNIT PROCESS GUIDE

INTRODUCTION

The Unit Process Guide consists of three parts. The first part concerns the 23 major unit processes and consists of Tables 3.1 and 3.2. The second part, concerning the 16 minor unit processes, consists of Tables 3.3 and 3.4. The third part of the Guide consists of Table 3.5 which was described in Chapter 2.

MAJOR UNIT PROCESSES

A brief definition of each of the 23 major unit processes is given in Chapter 4. Complete description of these processes was not attempted. The definitions used concern the products and processes as used in the tables, with specific product examples being used where applicable.

MINOR UNIT PROCESSES

Chapter 5 presents definitions of the 16 minor unit processes. Because some of these processes are less well known, a more detailed description has been included than for the major unit processes. Among these more detailed descriptions are the following: (1) the Ashai electrohydrodimerization process for continuous production of adiponitrile by reductive dimerization of acrylonitrile; (2) the Daicel process for epoxidation of propylene with peracetic acid to produce propylene and acetic acid; (3) the Stamicarbon BW process for production of caprolactam by oximation of cyclohexylamine, followed by a Beckmann rearrangement of the cyclohexanone oxime to caprolactam; and (4) the U.S. Industrial Chemicals process for the production of vinyl acetate by oxyacetylation from ethylene, acetic acid and oxygen.

DIRECTORY OF COMPANIES OWNING/LICENSING MANUFACTURING PROCESSES

There are 145 listings in Chapter 6 of companies, or major divisions of large companies, that own and/or license the processes for the products tabulated in Tables 3.2 and 3.4. Reference to product processes under license is given in Tables 3.2, 3.4 and 3.5 for a specific product or process.

Table 3.1 Major Unit Processes–Summary

Process	Compounds	Entries to Table
1. Alkylation	15	15
2. Amination by Ammonolysis	13	14
3. Ammoxidation	10	10
4. Carbonylation (oxo)	10	9
5. Condensation	55	56
6. Cracking (catalytic)	3	7
7. Dehydration	6	6
8. Dehydrogenation	15	20
9. Dehydrohalogenation	6	9
10. Esterification	24	24
11. Halogenation	54	60
12. Hydrodealkylation	3	5
13. Hydogenation	27	27
14. Hydrohalogenation	7	7
15. Hydrolysis (hydration)	28	31
16. Nitration	12	12
17. Oxidation	47	45
18. Oxyhalogenation	5	6
19. Phosgenation	3	3
20. Polymerization	34	38
21. Pyrolysis	20	18
22. Reforming (steam) - Water Gas Reaction	1	6
23. Sulfonation	11	9
TOTAL	410	437

Table 3.2 Major Unit Processes–Alkylation

Other Required Processes	Product	Feedstock	Owner of Process	No. of Licensed Commercial Plants[a]	Total No. of U.S. Commercial Plants[b]
Oxidation	Acetic acid	*n*-Butenes	Bayer AG	N	(10)
None	Alkylbenzenes (branched)	Benzene	Chevron Research	6	(3)
		Propylene tetramer	Conoco Chemicals	2	–
			Texaco Development Corp.	2	–
			Phillips Petroleum Co.	4	–
			UOP Process Div.	>5	
None	Alkylbenzenes (linear)	Benzene	Conoco Chemicals	2	(1)
		Linear olefins	Phillips Petroleum Co.	1	–
			UOP Process Div.	U	–
Dehydrogenation	Alkylbenzenes (linear)	Benzene	UOP Process Div.	4	(3)
		Linear paraffins	Hüls	1	–
None	Benzene, Xylenes	Toluene	UOP Process Div./Toray Ind. Inc.	1	(47)
None	*p-tert*-Butylphenol	Phenol	Hüls	1	(5)
		Isobutene			
None	Cumene	Benzene	Hüls	1	(13)
		Propylene	Union Carbide	2	–
			UOP Process Division	U	–
			Institut Francais du Petrole	N	–
None	Ethylbenzene	Benzene	Hüls	1	(17)
		Ethylene	Royal Dutch Shell	U	–
			Union Carbide	14	–
			UOP Process Div.	U	–

Table 3.2, continued

Other Required Processes	Product	Feedstock	Owner of Process	No. of Licensed Commercial Plants[a]	Total No. of U.S. Commercial Plants[b]
Dehydro-halogenation	N-Isopropyl-N′-phenyl-*p*-phenylenediamine[c]	*p*-Chloronitrobenzene Aniline	Sumitomo Chem. Co.	1	(3)
Hydrogenation		Acetone			
None	Lead alkyls	Ethyl chloride (alkyl chlorides)		–	(6)
None	*p*-Nonyl phenol	Phenol, Propylene trimer	Hüls	1	(13)
Acid cleavage Hydrolysis Oxidation	Phenol, Acetone	Benzene Propylene	Allied Chemical	U	(10)
Oxidation	Pyromellitic dianhydride	1,2,4-Trimethylbenzene (Pseudocumene)	–	–	(2)
Dehydrogenation	Styrene	Benzene Ethylene	Badger (Union Carbide-Cosden)	>20	(13)
			CdF-Chemie-Technip	7	–
			Monsanto	10	–
			Scientific Design	3	–
			Shell	U	–
None	2,4-Xylenol	*p*-Cresol Methyl Chloride	–	–	(5)

Major Unit Processes–Amination by Ammonolysis

None	Aniline	Phenol	Halcon Int.	1	(7)
None	Benzene sulfonamide	Benzene sulfonyl chloride	Diamond Shamrock Corp.	U	(0)
Condensation Hydrohalogenation	Choline chloride	Ethylene oxide Trimethylamine	UCB	1	(6)
None	*p*-Chlorobenzene sulfonamide	*p*-Chlorobenzene sulfonyl chloride	Daimond Shamrock Corp.	U	(2)
None	Dimethylformamide	Dimethylamine Methylformate	UCB	1	(3)
None	Ethanolamines	Ethylene oxide	Hüls	1	(5)
			Mitsui Toatsu	U	–
			Scientific Design	2	–
			Shell Development Co.	4	–
None	Ethylamines	Ethanol	Gulf Chemicals Co.	N	(5)
			Leonard Process Co.	4	–
None	Ethylenediamine	Ethylene dichloride	–	–	(3)
None	Ethylenediamine	Monoethanolamine	Leonard Process Co.	N	–
None	Hexamethylenediamine	Adipic acid	Zimmer/Beaunit	1	(6)
Condensation	Hexamethylenetetramine	Formaldehyde	Hooker	U	(6)
			Meissner	U	
None	Methylamines	Methanol	Leonard Process Co.	19	(6)
Condensation	2-Methyl-5-ethylpyridine (MEP) (5-ethyl-2-picoline)	Acetaldehyde	Montedison S.p.A.	1	(2)

Table 3.2, continued

Other Required Processes	Product	Feedstock	Owner of Process	No. of Licensed Commercial Plants[a]	Total No. of U.S. Commercial Plants[b]
Dehydration	Urea	Carbon dioxide	C & I Girdler	U	(49)
			Mitsui Toatsu	U	–
			Montedison S.p.A.	45	–
			Stamicarbon	103	–
			Marrovic/Technip	1	–
Major Unit Processes–Ammoxidation					
None	Acrylonitrile	Propylene	BP Chemicals International Ltd.	3	(5)
			Montedison S.p.A./UOP Process Division	2	–
			SNAM Progetti	1	–
			Standard Oil of Ohio/ Badger Co.	>45	–
None	Adiponitrile	Adipic acid	Montefibre S.p.A.	3	(7)
Halogenation Hydrogenation	Adiponitrile Hexamethylenediamine	Butadiene	–	–	–
None	Benzonitrile	Toluene	Mitsubishi Gas Chemical/ Badger Co.	N	(1)
Hydrogenation	Hexamethylenediamine	Adipic acid Ammonia	Zimmer/El Paso-Beaunit	1	(6)
None	Hydrogen cyanide	Methane	B. F. Goodrich	1	(11)
			Montedison S.p.A.	1	–
			Zimmer AG	1	–

None	Isophthalonitrile	*m*-Xylene	Mitsubishi Gas Chemical/ Badger Co.	2	(1)
None	Phthalonitrile	*o*-Xylene	Mitsubishi Gas Chemical/ Badger Co.	N	(0)
Condensation	Pyridine, *beta*-Picoline	Acetaldehyde Formaldehyde Methanol	BP Chemicals International Ltd.	N	(2)
None	Terephthalonitrile	*p*-Xylene	Mitsubishi Gas Chemical/ Badger Co.	N	(0)
		Major Unit Processes–Carbonylation (oxo)			
None	Acetic acid	Methanol Carbon monoxide	BASF AG	2	(10)
None	Acetic acid	Methanol Carbon monoxide	Monsanto	8	2
None	Alcohols (C_7-C_{13})	Olefins Carbon monoxide	Gulf Oil Chemicals Co. BASF AG	1 1	(7) –
Hydrogenation	*n*-Butanol *n*-Butyraldehyde	Propylene Carbon monoxide	BASF AG Ruhrchemie/FWH Farbwerke Hoechst	7 15	(7) –
None	Ethyl acrylate	Acetylene, Ethanol, Carbon monoxide	–	0	(5)
Hydrogenation	2-Ethyl hexanol *n*-Butyl alcohol	Propylene Carbon monoxide	Ruhrchemie AG and Ruhrchemie AG/ Rhone Progil SA	12 –	(5) –
	Isobutyraldehyde		Monsanto	N	–
Hydrogenation	Isobutyl alcohol	Propylene Carbon monoxide	–	–	(8)

Table 3.2, continued

Other Required Processes	Product	Feedstock	Owner of Process	No. of Licensed Commercial Plants[a]	Total No. of U.S. Commercial Plants[b]
Hydrolysis	Formic acid Sodium formate	Carbon monoxide Sodium hydroxide	Stauffer Chemical Co.	1	(2)
Condensation Hydrolysis	Formic acid	Carbon monoxide (methanol recycled)	Leonard Process Co., Inc.	N	–
Major Unit Processes–Condensation					
Pyrolysis	Acetic anhydride	Acetic acid	Wacker-Chemie	12+	(7)
None	Arsanilic acid (*p*-aminobenzenearsonic acid)	Aniline Arsenic acid	Sherwin-Williams	N	(3)
None	Benzene sulfonyl chloride	Benzene Chlorosulfonic acid	Diamond Shamrock Corp.	U	(2)
Dehydrogenation	Biphenyl (diphenyl)	Benzene	Nippon Steel	U	(8)
None	Bisphenol A	Acetone Phenol	Honshu	U	(6)
			Hooker Chemical Corp.	2	–
			Monsanto	1	–
			Union Carbide	2	–
			Rhone Progil	1	–
Amination by Ammonolysis Hydrohalogenation	Choline chloride	Ethylene oxide Trimethylamine	UCB	1	(6)

Hydrogenation	Crotonaldehyde, *n*-Butyl alcohol, *n*-Butyraldehyde	Acetaldehyde	BP Chemicals International Ltd.	U	(2)
			Kyowa Hakko	1	–
Halogenation	Dichlorodiphenyltrichloroethane [1,1,1-trichloro-2,2-*bis*(*p*-chlorophenyl) ethane] (DDT)	Acetaldehyde Monochlorobenzene	Diamond Shamrock Corp.	U	(1)
Halogenation	2,4-Dichlorophenoxyacetic acid (2,4-D)	Monochloroacetic acid	Diamond Shamrock Corp.	U	(11)
		Phenol	Mitsui Chemical	U	–
Dehydro-halogenation	2-(2,4-Dichlorophenoxy)propionic acid (2,4-DP)	α-Chloropropionic acid 2,4-Dichlorophenol			
None	4,4′-Dichlorophenylsulfone	Monochlorobenzene Sulfur trioxide	Diamond Shamrock Corp.	(1)	(0)
None	Diphenylamine	Aniline	Honshu	U	(3)
			Uniroyal	U	–
None	Diphenylguanidine	Aniline Cyanic acid	Sumitomo Chemical Co.	1	(1)
Hydrogenation	N,N′-Diphenylhydrazine (hydrazobenzene)	Nitrobenzene	–	–	(1)
Phosgenation	Diphenylmethane-4,4′-diisocyanate [(methylene *bis*(4-phenyl isocyanate)] (MDI)	Aniline Formaldehyde Phosgene	Hodogaya Chemical Co.	U	(4)
None	Ethyl acetate	Acetaldehyde	Hüls	1	(11)
			Wacher-Chemie	5	–
None	Ethylene carbonate	Ethylene oxide Carbon dioxide	Hüls	2	(1)
None	Ethylene glycol ethers	Alkyl alcohols Ethylene glycol	Shell Development Co.	3	(7)
None	Ethylene glycol monoethyl ether	Ethylene oxide Ethanol	Shell Development Co.	U	(5)

Table 3.2, continued

Other Required Processes	Product	Feedstock	Owner of Process	No. of Licensed Commercial Plants[a]	Total No. of U.S. Commercial Plants[b]
None	Ethyl ether (diethyl ether)	Ethanol	–	–	(6)
Hydrogenation	2-Ethylhexanol	Acetaldehyde	Kyowa Hakko	1	(5)
		Butyraldehyde	Rhone Progil	1	–
			Huls	2	–
			Kyowa Hakko	1	–
Halogenation	Ethyl parathion (Parathion)	O,O-Dimethyl phos-phorothionochloridate	Ruhrchemie/Rhone Progil	15	(2)
			Stauffer Chem. Co.	2	–
		Sodium nitrophenoxide			
Carbonylation Hydrolysis	Formic acid	Carbon monoxide (methanol recycled)	Leonard Process Co., Inc.	N	–
None	Heptenes	Butylenes	Institut Francais du Petrole	N	(2)
		Propylene	UOP Process Division	12	–
Amination by Ammonolysis	Hexamethylenetetramine	Ammonia	Hooker	U	(6)
		Formaldehyde	Meissner	U	–
None	Isophorone	Acetone	Veba-Chemie AG	U	(2)
Cracking	Isoprene (2-methyl-1,3-butadiene)	Propylene	Scientific Design	2	(6)
Dehydration Hydrogenation	Isoprene (2-methyl-1,3-butadiene)	Acetone Acetylene	SNAM Progetti	1	–
Cracking	Isoprene (2-methyl-1,3-butadiene)	Formaldehyde	Institut Francais du Petrole	1	–
		Isobutylene	Licensintorg (USSR)	2	–

Hydro-halogenation Pyrolysis	Isoprene (2-methyl-1,3-Butadiene)	Formaldehyde	Bayer	N	–
		Hydrogen chloride Isobutylene	Marathon Oil	N	–
Alkylation Dehydrohalogenation Hydrogenation	N-Isopropyl-N′-phenyl-*p*-phenylenediamine	Acetone Aniline *p*-Chloronitrobenzene	Sumitomo Chemical Co.	1	(3)
Pyrolysis	Melamine	Dicyandiamide	Produits Azole	U	(3)
Pyrolysis	Melamine	Urea	Montedison S.p.A. Nissan Chemical Industries	3 1	– –
Dehydration	Mesityl oxide	Acetone	BP Chemicals International Ltd. Texaco Development Corp.	3 U	(4) –
Hydrogenation	DL-Methionine	Acrolein Cyanic acid Methyl mercaptan	Sumitomo Chemical Co.	1	(2)
Hydrogenation	2-Methyl-2-butanol (*tert*-amyl alcohol)	Acetone Acetylene	SNAM Progetti	U	(2)
None	2-Methyl-3-butyn-2-ol	Acetone Acetylene	SNAM Progetti	1	(1)
Halogenation	2-Methyl-4-chlorophenoxy-acetic acid (MCPA)	*o*-Cresol Monochloroacetic acid	Mitsui Chemical	U	(2)
Dehydro-halogenation	2-(2-Methyl-4-chlorophenoxy)propionic acid, (MCPP)	α-Chloropropionic acid 4-Chloro-*o*-cresol	Lankro	U	(4)
Amination by Ammonolysis	2-Methyl-5-ethylpyridine (MEP) (5-ethyl-2-picoline)	Acetaldehyde Ammonia	Montedison S.p.A.	1	(2)

Table 3.2, continued

Other Required Processes	Product	Feedstock	Owner of Process	No. of Licensed Commercial Plants[a]	Total No. of U.S. Commercial Plants[b]
Dehydration Hydrogenation	Methyl isobutyl ketone	Acetone Hydrogen	BP Chemicals International Ltd. Texaco Development Corp.	3 2	5
Halogenation	Methyl parathion (0,0-dimethyl 0-*p*-nitrophenyl phosphorothioate)	0,0-Dimethyl phosphonochloridate Sodium *p*-nitrophenoxide	Stauffer Chemical Co.	1	(4)
None	4-Methyl-l-pentene	Propylene	BP Chemicals International Ltd.	U	(1)
None	N-Phenyl-2-naphthylamine	Aniline 2-Naphthol	Sumitomo Chemical Co.	1	(1)
Pyrolysis	Oxalic acid	Sodium formate	Stauffer Chemical Co.	1	(7)
Cannizzaro Reaction	Pentaerythritol	Acetaldehyde	Montedison S.p.A.	1	(4)
		Formaldehyde	Meissner	U	–
Dehydrogenation	*p*-Phenylphenol (4-hydroxydiphenyl)	Benzene Cyclohexanone	Union Carbide	N	(1)
Polymerization	Polyethylene terephthalate	Dimethyl terephthalate	Imperial Chemical Industries	11	(21)
		Ethylene glycol	Hoechst	U	–
			Inventa-Ems	4	–
			Zimmer AG	26	–
			Beaunit Corp.	U	–
None	*beta*-Propiolactone	Formaldehyde Ketene	B. F. Goodrich	2	(0)

None	Propylene carbonate	Carbon dioxide Propylene oxide	Hüls	2	(1)
Ammoxidation	Pyridine, *beta*-Picoline	Acetaldehyde Formaldehyde Methanol	BP Chemicals International Ltd.	N	(2)
Oxidation	Tetrahydrofuran, 2,3,4,5-tetracarboxylic dianhydride	Furan Maleic anhydride	Gelsenberg Chemie GmbH	N	(2)
Oxidation	Tetramethylthiuram disulfide (Thiram) [*bis*(dimethylthiocarbamoyl)disulfide]	Ammonia, Carbon disulfide Dimethylamine, Hydrogen peroxide	UCB	1	(5)
Halogenation	2,4,5-Trichlorophenoxyacetic acid (2,4,5-T)	Acetic acid, 2,4,5-Trichlorophenol	Diamond Shamrock Corp.	U	(8)
None	Zineb (zinc ethylenebisdithiocarbamate)	Carbon disulfide Ethylenediamine Zinc sulfate	UCB	1	(4)
None	Ziram (zinc dimethyldithiocarbamate)	Carbon disulfide Dimethylamine Zinc sulfate	UCB	1	(9)

Major Unit Processes–Cracking, Catalytic

None	Isobutylene (isobutene)	Isobutane	Coastal States Petroleum Co.-Foster Wheeler	1	(1)
			Houdry Div. Air Prod. & Chem.	U	–
			Philips Petroleum Co.	2	–
			Total Compagnie Francaise de Raffinage	3	–
			UOP Process Div.	N	–

Table 3.2, continued

Other Required Processes	Product	Feedstock	Owner of Process	No. of Licensed Commercial Plants[a]	Total No. of U.S. Commercial Plants[b]
Condensation	Isoprene	Propylene	Scientific Design	U	–
None	Isoprene	Amylenes	Shell	2	–
Condensation	Isoprene	Isobutylene	Institut Francais du Petrole	1	–
		Formaldehyde	Licenstorg (USSR)	2	–
Halogenation	Perchloroethylene and trichloroethylene	Ethylene dichloride	Diamond Shamrock Corp.	U	(11)
			Rhone Progil	1	–
			Scientific Design	5	–
Halogenation	Perchloroethylene and trichloroethylene	Acetylene	Wacker-Chemie	3	–
Halogenation Oxyhalogenation	Perchloroethylene and trichloroethylene	Any C_2 chlorocarbon mixture	PPG Ind.	U	–
Halogenation	Vinyl chloride monomer (VCM)	Acetylene	Monochem, Inc.	–	(1)
Halogenation	Vinyl chloride monomer (VCM)	Ethylene	–	–	(2)
Dehydro-halogenation	Vinyl chloride monomer (VCM)	Naphtha	Kureha-Chiyoda	3	–
Halogenation Oxyhalogenation			Nippon Zeon Co. Ltd.	1	–
Halogenation Oxyhalogenation	Vinyl chloride monomer (VCM)	Ethylene	B. F. Goodrich Chemical	26	–

Major Unit Process–Dehydration

None	Ethylene	Ethyl alcohol	Air Reduction Co.	2	–
Condensation Hydrogenation	Isoprene(2-methyl-1,3-Butadiene)	Acetone Acetylene	SNAM Progetti	1	(6)
Condensation	Mesityl oxide (isopropylidene acetone)	Acetone	BP Chemicals International Ltd.	3	(5)
			Texaco Development Corp.	U	–
Condensation Hydrogenation	Methyl isobutyl ketone	Acetone Hydrogen	BP Chemicals International Ltd.	3	(5)
			Texaco Development Corp.	U	–
None	Morpholine	Diethanolamine	Hüls	1	(1)
Animation by Ammonolysis	Urea	Ammonia Carbon dioxide	C & T Girdler	U	(49)
			Mavrovic Technip	1	–
			Mitsui Toatsu Chemicals, Inc.	61	–
			Montedison S.p.A.	45	–
			SNAM Progetti	10	–
			Stamicarbon	150	–

Major Unit Process–Dehydrogenation

None	Acetaldehyde	Ethyl alcohol	BP Chemicals International Ltd.	4	(2)
			Veba-Chemie AG	2	–
None	Acetone	Isopropyl alcohol	BP Chemicals International Ltd.	3	(7)
			Deutsche Texaco AG	2	–
Alkylation	Alkyl benzenes, linear	Benzene Linear paraffins	UOP Process Division	4	(3)
			Huls	1	–
Condensation	Biphenyl (diphenyl)	Benzene	Nippon Steel Chemical	U	(8)
None[d]	Butadiene	*n*-Butane	Houdry Div. Air Products & Chem.	14	(2)
			Philips Petroleum Co.	U	–

Table 3.2, continued

Other Required Processes	Product	Feedstock	Owner of Process	No. of Licensed Commercial Plants[a]	Total No. of U.S. Commercial Plants[b]
None[d]	Butadiene	Butene-1 Butene-2	Houdry Div. Air Products & Chem.	3	(2)
None[e]	Butadiene	Butene-1 Butene-2	Philips Petroleum Co. BP Chemicals International Ltd.	U N	– –
None	Cyclohexanone	Cyclohexanol Cyclohexane mixtures	Zimmer AG Institute Francais du Petrole	4 N	– –
None	α-Isoamylene (3-methyl-l-butene)	Isopentane	Houdry Div., Air Prod. & Chem.	U	(1)
None	Isoprene (2-methyl-1,3-butadiene)	Isopentane	Houdry Div., Air Prod. & Chem.	U	–
None	Isoprene (2-methyl-1,3-butadiene)	Tertiary amylenes	Shell Development Co.	4	(1)
Hydrolysis	Methyl ethyl ketone	Butene-1 Butene-2	Texaco Development Corp.	U	(7)
None	Methyl ethyl ketone	*sec*-Butyl alcohol	BP Chemicals International Ltd. Scientific Design-Maruzen Oil	1 1	4 –
Oxidation	Phenol	Cyclohexane	Scientific Design Institute Francais du Petrole	1 N	(17) –
Condensation	*p*-Phenylphenol (4-hydroxydiphenyl)	Benzene Cyclohexanone	Union Carbide	N	(1)
None	Piperylene (1,3-pentadiene)	*n*-Pentene	Houdry Div., Air Prod. & Chem.	U	(2)
None	Propylene	Propane	Houdry Div., Air Prod. & Chem. UOP Process Div.	U N	(61) –

Alkylation	Styrene	Benzene	Badger (Union Carbide and Cosden)	>20	(13)
		Ethylene	CdF-Chemie/Technip	7	–
			Monsanto	10	–
			Shell Development Co.	U	–
			Scientific Design	3	–
None	Styrene	Ethylbenzene	Badger (Union Carbide and Cosden)	16	–
			Hüls	2	–
			Monsanto	2	–
			Scientific Design	3	–
			Shell Development Co.	U	–
			UOP Process Div.	U	–
None	Xylenes, mixed	Naphtha	Houdry Div., Air Products & Chem.	U	(29)

Major Unit Processes–Dehydrohalogenation

Condensation	2-(2,4-Dichlorophenoxy) propionic acid (2,4-DP)	α-Chloropropionic acid (2,4-Dichlorophenol)	Lankro	U	(2)
Alkylation Condensation Hydrogenation	N-Isopropyl-N′-phenyl-*p*-phenylenediamine	Acetone Aniline *p*-Chloronitrobenzene	Sumitomo Chemical Co.	1	(3)
Condensation	2-(2-Methyl-4-chlorophenoxy)propionic acid (MCPP)	α-Chloropropionic acid; 4-Chlorocresol	Lankro	U	(4)
Phosgenation Polymerization	Polycarbonate resins	Bisphenol A Phosgene	Identsu Kosan	1	(3)

Table 3.2, continued

Other Required Processes	Product	Feedstock	Owner of Process	No. of Licensed Commercial Plants[a]	Total No. of U.S. Commercial Plants[b]
Halogenation	Vinyl chloride monomer (VCM)	Acetylene	Monochem, Inc.	U	U
Halogenation	Vinyl chloride monomer (VCM)	Ethane	Transcat/Lummus Co.	U	–
Oxyhalogenation		Chlorine			
Halogenation	Vinyl chloride monomer (VCM)	Ethylene	B. F. Goodrich	31	(14)
Oxyhalogenation		Chlorine	Montedison S.p.A.	1	–
			Monsanto/Scientific Design	6	–
			P.P.G. Industries	U	–
			Rhone Progil	2	–
			Stauffer Chemical Co.	22	–
			Union Carbide	1	–
None	Vinyl chloride monomer (VCM)	Ethylene dichloride	BF Goodrich/Hoechst	20	–
	Hydrogen chloride		Farbwerke Hoechst	28	–
			Kureha Chemical Co.	1	–
			Rhone Progil	2	–
			Solvay-ICI	U	–
			Stauffer Chemical Co.	12	–
			Wacker-Chemie	U	–
Halogenation	Vinyl chloride monomer (VCM)	Naphtha	Kureha-Chiyoda	3	–
Oxyhalogenation		Chlorine	Nippon Zeon Co. Ltd.	1	–
Halogenation	Vinylidene chloride (1,1-dichloroethylene)	Vinyl chloride	–	–	(3)

Major Unit Processes–Esterification

Oxidation	Acrylic acid and acrylate esters	Propylene Alcohols	BP Chemicals International Ltd.	1	(9)
			Mitsubishi Petrochemical, Ltd.	1	–
			Nippon Shokubai Kagaku Kogyo Co., Ltd.	1	–
			Toyo Soda Manufacturing Co., Inc.	N	–
None	*n*-Butyl acetate	Acetic acid *n*-Butyl alcohol	BP Chemicals International, Inc.	U	(5)
None	*n*-Butyl benzyl phthalate	Benzyl alcohol *n*-Butyl alcohol Phthalic anhydride	–	–	(1)
None	Di-*n*-butyl phthalate	*n*-Butyl alcohol Phthalic anhydride	BP Chemicals International, Ltd.	U	(10)
			Rhone Progil	1	–
None	Diethyl phthalate	Ethyl alcohol Phthalic anhydride	BP Chemicals International, Ltd.	2	(4)
None	Diheptyl phthalate	Heptyl alcohol Phthalic anhydride	BP Chemicals International, Ltd.	2	(1)
			Nihon Yuki	U	–
None	Diisodecyl phthalate	Isodecyl alcohol Phthalic anhydride	–	–	(8)
None	Dimethyl phthalate	Methyl alcohol Phthalic anhydride	–	–	(5)
None	Dimethyl terephthalate	Methyl alcohol	Eastman Kodak Co.-Tennessee Eastman Division	2	(7)
		Terephthalic acid	Mitsui Petrochemical Industries, Ltd.	2	–
			Standard Oil of Indiana	14	–

Table 3.2, continued

Other Required Processes	Product	Feedstock	Owner of Process	No. of Licensed Commercial Plants[a]	Total No. of U.S. Commercial Plants[b]
Oxidation	Dimethyl terephthalate	Methyl alcohol	Dynamit Nobel AG	25	–
		p-Xylene	Katzchmann/Simon Carres	13	–
			C. W. Witten	14	–
None	Di-*n*-octyl phthalate	Phthalic anhydride	BP Chemicals International Ltd.	2	(0)
		n-Octyl alcohol	Lankro	4	–
			Nihon Yuki	U	–
			Rhone Progil	6	–
None	Dioctylphthalate (2-ethylhexyl phthalate)	2-Ethylhexyl alcohol	BP Chemicals International Ltd.	2	(10)
		Phthalic anhydride	Lankro	4	–
			Pfaudler	4	–
			Rhone Progil	6	–
None	Ethyl acetate	Acetic acid Ethyl Alcohol	BP Chemicals International Ltd.	U	(11)
Pyrolysis	Ethyl acetoacetate	Acetic acid Ethyl alcohol	Wacker-Chemie	U	2
Oxidation	Ethyl acrylate	Ethyl alcohol Propylene		–	(5)
None	Isopropyl acetate	Acetic acid Isopropyl alcohol	BP Chemicals International Ltd.	U	(3)
None	Methyl acetate	Acetic acid Methyl alcohol	BP Chemicals International Ltd.	U	(8)
Pyrolysis	Methyl acetoacetate	Acetic acid Isopropyl alcohol	Wacker-Chemie	U	(1)

Hydrocyanation Hydrolysis Sulfonation	Methyl methacrylate	Acetone Hydrogen cyanide Methyl alcohol	Montedison S.p.A. Sumitomo Chemical Co.	1 1	(4) –
Oxidation	Methyl methacrylate	Isobutylene Methyl alcohol	Gulf Oil Chemicals Co.	N	–
Carboxylation	*p*-Oxybenzoic acid and *p*-oxybenzoic butyrate	Butyl alcohol Carbon dioxide Phenol	Mitsui Chemical	U	–
Polymerization	Polyethylene terephthalate	Ethylene glycol Terephthalic acid	Hoechst ICI Inventa-Ems Toray Industries, Inc. Zimmer AG	U U 12 U 36	(21) – – – –
None	Triacetate polymer (cellulose triacetate)	Acetic acid Cellulose	Montefibre S.p.A.	2	–
None	Tributyrin (glyceryl tributyrate)	*n*-Butyric acid Glycerol	Nihon Yuki	N	–

Major Unit Processes–Halogenation

Ammoxidation Hydrogenation	Adiponitrile, hexamethylenediamine	Butadiene	–	–	(7)
None	Allyl chloride (3-chloroprene)	Propylene	–	–	(5)
None	Benzyl chloride (α-chlorotoluene)	Toluene	–	–	(4)
None	Bromoform	Acetone Bromine	–	–	(2)
None	Carbon tetrachloride	Carbon disulfide	Diamond Shamrock Corp. Stauffer Chemical Co.	U 4	(2) –

Table 3.2, continued

Other Required Processes	Product	Feedstock	Owner of Process	No. of Licensed Commercial Plants[a]	Total No. of U.S. Commercial Plants[b]
None	Carbon tetrachloride, Perchloroethylene	Methane	Rhone Progil	3	(4)
			Scientific Design Co.	5	–
			Stauffer Chemical Co.	5	–
None	Carbon tetrachloride, Perchloroethylene	Propane/Propylene	Rhone Progil	U	(4)
			Scientific Design Co.	5	–
			Stauffer Chemical Co.	1	–
None	Chloral (trichloroacetaldehyde)	Acetaldehyde	Diamond Shamrock Corp.	1	(2)
None	Chloroacetic acid	Acetic acid	–	–	(3)
None	Chlorobenzene, mono (MCB)	Benzene	–	–	(8)
None	*p*-Chloro-*m*-cresol (2-chloro-5-hydroxy-toluene, 4-chloro-3-methylphenol)	*m*-Cresol	–	–	(3)
None	Chloromethanes/Carbon tetrachloride, Chloroform, Methyl chloride, Methylene chloride	Methane	Diamond Shamrock Corp.	U	(6)
			Scientific Design Co.	1	–
			Stauffer Chemical Co.	1	–
None	2-Chloronaphthalene	β-Naphthol	–	–	(1)
None	*o*-Chlorophenol *p*-Chlorophenol	Phenol	–	–	(2)
None	Choropicrin (trichloronitromethane)	Picric acid	Mitsui Chemical	U	(5)
None	Chloroprene (2-chlorobutadiene-1,3)	Butadiene	BP Chemicals International Ltd.	3	(3)

Pyrolysis	Cyanuric acid, Sodium dichloroisocyanurate, trichloroisocyanuric acid	Caustic soda, chlorine, urea	Shikoku Kasei	3	–
None	*o*-Dichlorobenzene (1,2-dichlorobenzene) *p*-Dichlorobenzene (1,4-dichlorobenzene)	Benzene	Mitsui Chemical	1	(7)
None	*m*-Dichlorobenzene (1,3-dichlorobenzene)	Benzene	–	–	(2)
Condensation	Dichlorodiphenyltrichloroethane [1,1,1-Trichloro-2,2-bis(*p*-chlorophenyl) ethane] (DDT)	Acetaldehyde Monochlorobenzene	Diamond Shamrock Corp.	U	(1)
None	2,4-Dichlorophenol	Phenol, 4-chlorophenol	–	–	(3)
None	2,4-Dichlorophenoxyacetic acid (2,4-D)	Monochloroacetic acid Phenol	Diamond Shamrock Corp. Nissan Chemical Industries	U U	(11) –
None	1,2-Dichloropropane	Propylene	–	–	(5)
None	1,3-Dichloropropane (1,3-D)	Allyl chloride	–	–	(1)
None	Ethylene dibromide (1,2-dibromoethane)	Ethylene	–	–	(4)
None	Ethylene dichloride (1,2-dichloroethane)	Ethylene	Diamond Shamrock Corp.	U	(17)
			BF Goodrich	6	–
			Hüls	1	–
			Monsanto	5	–
			Nippon Zeon Co., Ltd.	U	–
			Rhone Progil	2	–
			Solvay-Imperial Chemical Ind.	U	–
			Stauffer Chemical Co.	18	–
			Toyo Soda	U	–
			Toyo Koatsu	N	–
			Wacker-Chemie	U	–
Condensation	Ethyl parathion (parathion, 0,0-diethyl 0-*p*-nitrophenyl phosphorothioate)	0,0-Dimethyl phosphorothionochloridate Sodium nitrophenoxide	Stauffer Chemical Co.	2	(2)

Table 3.2, continued

Other Required Processes	Product	Feedstock	Owner of Process	No. of Licensed Commercial Plants[a]	Total No. of U.S. Commercial Plants[b]
None	Fluorocarbons	Carbon tetrachloride	Daikin Kogyo Co., Ltd.	U	(16)
		Hydrofluoric acid	Rhone Progil	2	–
None	Fluorocarbons	Hydrofluoric acid Methane	Montedison S.p.A.	1	–
Chloro-hydrination Hydrolysis	Glycerine (glycerol)	Propylene→allyl chloride→ epichlorohydrin	–	–	(2)
None	Hexachlorobenzene (HCB) (perchlorobenzene)	Benzene	Diamond Shamrock Corp.	U	(2)
			Mitsui Chemical	U	–
			Stauffer Chemical Co.	2	–
None	Hexachlorobutadiene	Butadiene	–	–	(3)
Pyrolysis	Hexachlorocyclopentadiene (perchlorocyclopentadiene)	Pentane	–	–	(3)
None	Hexachloroethane	Tetrachloroethane	–	–	(1)
None	Methyl chloride (chloromethane) (methylene chloride, chloroform are coproducts)	Methane	–	–	(2)
Condensation	2-Methyl-4-chlorophenoxyacetic acid (MCPA)	*o*-Cresol Monochloroacetic acid	Mitsui Chemical	U	(2)

None	Methylene chloride (dichloromethane) (chloroform coproduct)	Methyl chloride	Diamond Shamrock Corp.	U	(5)
			Mitsui Chemical	U	–
			Stauffer Chemical Co.	2	–
None	Methylene chloride (methyl chloride, chloroform are coproducts)	Methane	–	–	(1)
None	Methylene chloride (methyl chloride, chloroform are coproducts)	Methane and methanol	–	–	(2)
Condensation	Methyl parathion (0,0-dimethyl 0-*p*-nitrophenyl phosphorothioate) (MPT)	0,0-Dimethyl phosphorothionochloridate Sodium *p*-nitrophenoxide	Stauffer Chemical Co.	1	(4)
None	Monochloroacetic acid	Acetic acid	Mitsui Chemical	U	(3)
			Uniroyal	U	–
None	Pentachlorophenol	Phenol	–	–	(5)
Cracking (catalytic)	Perchloroethylene (tetrachloroethylene) (trichloroethylene–coproduct)	Ethylene dichloride	Diamond Shamrock Corp.	U	(11)
			Rhone Progil	1	–
			Scientific Design	5	–
Cracking (catalytic)	Perchloroethylene (tetrachloroethylene) (trichloroethylene–coproduct)	Acetylene	Wacker-Chemie	3	–
Cracking (catalytic) Oxyhalogenation	Perchloroethylene (tetrachloroethylene) (trichloroethylene–coproduct)	Any C_2 chlorocarbon mixture	PPG Industries	U	–
None	Phosgene	Carbon monoxide	Crawford & Russell	14	(19)
			Stauffer Chemical Co.	5	–
			Zimmer AG	2	–
Chloro-hydrination Hydrolysis	Propylene oxide	Propylene *tert*-Butyl alcohol (recycled) Chlorine	Lummus	N	(0)

Table 3.2, continued

Other Required Processes	Product	Feedstock	Owner of Process	No. of Licensed Commercial Plants[a]	Total No. of U.S. Commercial Plants[b]
None	1,1,2,2-Tetrachloroethane	Ethane	–	–	(1)
None	1,2,4-Trichlorobenzene	1,2-Dichlorobenzene	–	–	(4)
None	1,1,1-Trichloroethane (methyl chloroform)	Ethylene	Detrex/Scientific Design	N	(3)
Hydro-halogenation	1,1,1-Trichloroethane (methyl chloroform)	Vinyl chloride	–	–	(3)
None	1,1,2-Trichloroethane	Ethylene	–	–	(1)
Hydrolysis	2,4,5-Trichlorophenol	Benzene Methanol Sodium hydroxide	–	–	(2)
Condensation	2,4,5-Trichlorophenoxyacetic acid (2,4,5-T)	Acetic acid, 2,4,5-Trichlorophenol	Diamond Shamrock Corp.	U	(8)
Dehydro-halogenation	Vinyl chloride monomer (VCM)	Acetylene	Monochem, Inc.	U	(1)
Dehydro-halogenation Oxyhalogenation	Vinyl chloride monomer (VCM)	Ethane Chlorine	Transcat/The Lummus Co.	U	
Dehydro-halogenation Oxyhalogenation	Vinyl chloride monomer (VCM)	Ethylene Chlorine	BF Goodrich Montedison S.p.A. Monsanto/Scientific Design P.P.G. Industries	31 1 6 U	(14)

			Rhone Progil	2	–
			Stauffer Chemical Co.	22	–
			Union Carbide	1	–
Dehydro-halogenation	Vinyl chloride monomer (VCM)	Naphtha	Kureha-Chiyoda	3	–
Oxyhalogenation		Chlorine	Nippon Zeon Co., Ltd.	1	–
Dehydro-halogenation	Vinylidene chloride (1,1-dichloroethylene)	Vinyl Chloride			

Major Unit Processes–Hydrodealkylation

None	Benzene Light Hydrocarbons	Aromatic mixtures Hydrogen	BASF/Veba Chemie AG	1	–
			British Gas Corp.	>20	–
			Gulf Oil Chemicals Co.	U	–
			Houdry Div. Air Products & Chem.	18	–
			Hydrocarbon Research Inc./ ARCO	12	–
			UOP Process Div.	>20	–
None	Benzene	Coke oven light oil	Houdry Div. Air Prod. & Chem.	4	–
None	Benzene	Toluene	British Gas Oil Co.	N	–
			Gulf R&D Co.	U	–
			Houdry Div. Air Prod. & Chem.	8	–
			Monsanto	U	–
			Phillips Petroleum Co.	1	–
			UOP Process Div.	2	–
None	Benzene	Xylenes (mixed)	Gulf R&D Co.	U	–
			Houdry Div. Air Prod. & Chem.	2	–
			UOP Process Div.	2	–

Table 3.2, continued

Other Required Processes	Product	Feedstock	Owner of Process	No. of Licensed Commercial Plants[a]	Total No. of U.S. Commercial Plants[b]
None	Naphthalene	Alkylnaphthalenes	Monsanto	U	–
			Houdry Div. Air Prod. & Chem.	2	–
			UOP Process Div.	2	–
Major Unit Processes–Hydrogenation					
Ammoxidation Halogenation	Adiponitrile	Butadiene	–	–	(7)
Hydrolysis Sulfonation	*m*-Aminophenol	Nitrobenzene	Sumitomo Chemical Co.	1	0
Acid Rearrangement	*p*-Aminophenol	Nitrobenzene	–	–	(4)
None	Aniline	Nitrobenzene	Lonza/First Chemical Corp.	U	(7)
			Mitsui Chemical	1	–
			Sumitomo Chemical Co.	1	–
None	1,3-*bis*(aminomethyl) cyclohexane	Isophthalonitrile	Mitsubishi Gas Chemical Co./Badger	1	(0)
Carbonylation (oxo)	*n*-Butanol/*n*-Butyraldehyde	Propylene Carbon monoxide	BASF AG	2	(7)
Acid Rearrangement Oxidation	Caprolactam	Toluene, Ammonia	SNIA Viscosa	U	(2)

Condensation Dehydration	Crotonaldehyde *n*-Butyraldehyde *n*-Butyl alcohol	Acetaldehyde	BP Chemicals International	U	(2)
			Kyowa Hakko	1	–
None	Cyclohexane	Benzene	Atlantic Richfield	2	(11)
			Houdry Div. Air Prod. & Chem.	2	–
			Institut Francais du Petrole	12	–
			Lummus	N	–
			Phillips Petroleum Co.	4	–
			Scientific Design	1	–
			Sinclair-Engelhard Ind.	U	–
			Stamicarbon	5	–
			Texaco Development Corp.	U	–
			Toray Ind. Inc., Hytoray	1	–
			UOP Process Div., Hydrar	1	–
None	Cyclohexanol	Phenol	Stamicarbon	2	2
None	Cyclohexylamine	Aniline	Abbott-Englehard Ind.	U	(4)
None	Cyclohexylamine	Nitrobenzene	Englehard Ind.	U	–
Benzidine Rearrangement	3,3′-Dichlorobenzidine dihydrochloride	1-Chloro-2-nitro-benzene	–	–	(2)
Carbonylation (oxo)	2-Ethylhexanol *n*-Butanol Isobutyraldehyde	Carbon monoxide Propylene	BASF AG	6	(5)
			Ruhrchemie AG		
			Ruhrchemie AG/Rhone Poulenc SA	12	–
			Monsanto	N	–
Condensation	2-Ethylhexanol	Acetaldehyde	Kyowa Hakko	1	(U)
			Rhone Progil	1	–
None	Hexamethylenediamine	Adiponitrile	Montefibre S.p.A.	1	(6)
Ammoxidation	Hexamethylenediamine	Adipic acid Ammonia	Zimmer/El Paso-Beaunit	1	–

Table 3.2, continued

Other Required Processes	Product	Feedstock	Owner of Process	No. of Licensed Commercial Plants[a]	Total No. of U.S. Commercial Plants[b]
Condensation	Hydrazobenzene *sym*-N,N′-diphenylhydrazine)	Nitrobenzene	–	–	1
Carbonylation (oxo)	Isobutyl alcohol	Propylene	–	–	(8)
Condensation Dehydration	Isoprene (2-methyl-1,3-butadiene)	Acetone Acetylene	SNAM Progetti	1	(6)
Alkylation Condensation Dehydrogenation	N-Isopropyl-N′-phenyl-*p*-phenylenediamine	Acetone Aniline *p*-Chloronitrobenzene	Sumitomo Chemical Co.	1	(3)
Condensation	DL-Methionine	Acrolein Cyanic acid Methyl mercaptan	Sumitomo Chemical Co.	1	(2)
Condensation	2-Methyl-2-butanol	Acetone Acetylene	SNAM Progetti	U	(2)
Condensation Dehydration	Methyl isobutyl ketone	Acetone	BP Chemicals International Ltd. Texaco Development Co.	3 U	(5) –
None	Sorbitol (1,2,3,4,5,6-hexanehexol)	Corn sugar or Corn syrup	Engelhard Ind.	U	(6)
Nitration Phosgenation	Toluene diisocyanate (TDI) (80/20-2,4-2,6-TDI)	Phosgene Toluene	Allied Chemical FMC Nippon Soda-Nissan Sumitomo Chemical Co.	U N U 1	(10) – – –

None	*m*-Xylenediamine	Isophthalonitrile	Mitsubishi Gas Chemical/Badger	1	1
Major Unit Processes–Hydrohalogenation					
Amination by Ammonolysis, Condensation	Choline chloride	Ethylene oxide Trimethylamine	UCB	1	(6)
None	Ethyl chloride	Ethylene	Stauffer	1	(7)
Condensation Pyrolysis	Isoprene	Formaldehyde	Bayer	N	(6)
		Isobutylene	Marathon Oil	N	–
None	Methyl bromide (bromomethane)	Methanol	–	–	(3)
None	Methyl chloride (chloromethane)	Methanol	Diamond Shamrock Corp.	1	(10)
			Stauffer Chemical Co.	2	–
Halogenation	1,1,1-Trichloroethane (methyl chloroform)	Vinyl chloride	–	–	(3)
None	Vinyl chloride monomer	Acetylene	Blaw Knox	1	1
			Crawford & Russell	2	–
			Huls	2	–
			Mitsui Chemical	U	–
			Scientific Design	2	–
			Wacker-Chemie	4	–
Major Unit Processes–Hydrolysis [Hydration]					
None	Acetic acid, Methanol	Methyl acetate	Wacher-Chemie	N	(10)
None	Acrylamide	Acrylonitrile	Sumitomo Chemical Co.	1	(4)
Sulfonation	Alcohols, mixed linear sulfated -ammonium salt	Fatty alcohols, sulfur trioxide	Chemithon	U	(9)
		Fatty alcohols, sulfur trioxide	MoDo Kemi AB	5	(2)

Table 3.2, continued

Other Required Processes	Product	Feedstock	Owner of Process	No. of Licensed Commercial Plants[a]	Total No. of U.S. Commercial Plants[b]
	-sodium salt	Fatty alcohols, sulfur trioxide	–	–	(4)
	-Triethanolamine salt	Fatty alcohols, sulfur trioxide	–	–	(2)
	-unspecified salts	Fatty alcohols, sulfur trioxide	–	–	(1)
Sulfonation	Alkylbenzene sulfonates	Alkylbenzenes, sulfur trioxide	Chemithon	U	(10)
			Hüls	1	–
			MoDo Kemi AB	5	–
Hydrogenation Sulfonation	*m*-Aminophenol	Nitrobenzene	Sumitomo Chemical Co.	1	(0)
Sulfonation	*sec*-Butyl alcohol	Butylene	Scientific Design/ Maruzen Oil Co.	1	(4)
None	*sec*-Butyl alcohol	Butene-1/butene-2	Texaco Development Corp.	U	–
Chloro-hydrination	Epichlorohydrin	Allyl chloride Hypochlorous acid	Union Carbide	1	(3)
None	Ethyl alcohol	Ethylene	National Distillers	1	(6)
			Shell	5	–
			Veba-Chemie AG	5	–
			Union Carbide	1	–
None	Ethyl alcohol, Ethyl ether	Ethylene	Vulcan-Cincinnati	3	–
			U.S. Industrial Chem. Co.	–	1

None	Ethylene glycol	Ethylene oxide	Hüls	4	(15)
			Japan Catalytic Chemical Ind. Co.	60	–
			Scientific Design	1	–
			Shell	22	–
Oxidation	Ethylene oxide, Ethylene glycol	Ethylene	Hüls	4	–
			Japan Catalytic Chem. Ind. Co.	6	–
			Nippon Shokubai Kagako Kogyo Co. Ltd.	3	–
			Scientific Design (ethylene oxide only)	49	–
			Shell	30	–
			SNAM Progetti S.p.A.	3	–
			Union Carbide	2	–
Carbonylation	Formic acid, Sodium formate	Carbon monoxide Sodium hydroxide	Stauffer Chemical Co.	1	(2)
Carbonylation Condensation	Formic acid	Carbon monoxide (methanol–recycled)	The Leonard Process Co, Inc.	N	–
Epoxidation	Glycerine, Acetic acid	Allyl alcohol, Peracetic acid	Daicel, Ltd.	N	(1)
Chloro-hydrination Halogenation Hydrolysis	Glycerine (glycerol)	Propylene→allylchlor-ide→epichlorohydrin	–	–	(2)
Sulfonation	Isopropyl alcohol	Propylene	BP Chemicals International Ltd.	4	(5)
			M. W. Kellog	N	–
			Texaco Development Corp.	U	–

Table 3.2, continued

Other Required Processes	Product	Feedstock	Owner of Process	No. of Licensed Commercial Plants[a]	Total No. of U.S. Commercial Plants[b]
None[f]	Isopropyl alcohol	Propylene	Veba-Chemie AG	5	–
			Deutsche Texaco AG	1	–
			Tokuyama Soda Co., Ltd.	1	–
None	Maleic acid (*cis*-1,2-ethylenedicarboxylic acid	Maleic anhydride	Bowmanns Chemical	1	–
Dehydrogenation	Methyl ethyl ketone	Butene-1/Butene-2	Texaco Development Corp.	U	(7)
Esterification Hydrocyanation Sulfonation	Methyl methacrylate monomer	Acetone Hydrogen cyanide Methanol	Montedison S.p.A. Sumitomo Chem. Co.	1 1	(4) –
Sulfonation	2-Naphthol	Naphthalene	Mitsui Chemical	U	(1)
None	Pentachlorophenol	Hexachlorobenzene	Diamond Shamrock Corp.	U	(5)
None	Phenol	Monochlorobenzene	Hooker Chemical Co.	3	(17)
Sulfonation	Phenol	Benzene, Sulfuric acid	Mitsui Chemical	U	1
Oxyhalogenation	Phenol	Benzene, Hydrogen chloride	Hooker Chemical Co. Union Carbide	3 1	– –
Acid cleavage Alkylation Oxidation	Phenol, Acetone	Benzene Propylene	Allied Chemical	U	–
None	Propylene glycol (1,2-dihydroxypropane)	Propylene oxide	Hüls/Bayer Mitsui Chemical	2 U	(6) –

Chlorohydrin-ation	Propylene oxide	Propylene	Lummus	N	(0)
Halogenation		*tert*-Butyl alcohol (recycled) Chlorine			
Chloro-hydrination	Propylene oxide	Propylene Chlorine solution	BASF AG Hüls/Bayer Union Carbide	1 2 >1	(6) – –
Halogenation	2,4,5-Trichlorophenol	Benzene Methanol Sodium hydroxide	–	–	(2)

Major Unit Processes–Nitration

Amination by Reduction	*o*-Aminophenol	Phenol	–	–	(2)
None	4,6-Dinitro-*o*-cresol	Cresol	–	–	(1)
None	2,4-Dinitrophenol	Phenol	–	–	(2)
None	2,4-Dinitrotoluene	Toluene	Biazzi	U	(4)
None	2,4-(and 2,6)-Dinitrotoluene	Toluene	Meissner Sumitomo Chemical	U 1	(4) –
None	Nitrobenzene	Benzene	Biazzi-Chemico Meissner Sumitomo Chemical Co. Uniroyal	U U 1 U	(8) – – –
None	*o*-Nitrophenol	Phenol	–	–	(2)
None	*p*-Nitrophenol	Phenol	–	–	(4)
None	*m*-Nitrotoluene	Toluene	Biazzi Meissner	U U	(1) –

Table 3.2, continued

Other Required Processes	Product	Feedstock	Owner of Process	No. of Licensed Commercial Plants[a]	Total No. of U.S. Commercial Plants[b]
None	*o*-Nitrotoluene	Toluene	Biazzi	U	(2)
			Meissner	U	–
None	*p*-Nitrotoluene	Toluene	Biazzi	U	(2)
			Meissner	U	–
Hydrogenation	Toluene diisocyanate (TDI)	Phosgene	Allied Chemical	U	(10)
Phosgenation	(80/20 2,4=2,6-TDI)	Toluene	FMC Corp.	N	–
			Nippon Soda-Nissan	U	–
			Sumitomo Chemical Co.	1	–
Major Unit Processes–Oxidation					
None	Acetaldehyde	Ethylene	Aldehyd GmbH	25	(5)
Alkylation	Acetic acid	*n*-Butenes	Bayer AG	N	(10)
None	Acetic acid	Light naphtha	BP Chemicals International Ltd.	6	–
None	Acetic acid, Peracetic acid	Acetaldehyde	Wacker-Chemie	1	–
None	Acetic acid, Terephthalic acid	*n*-Butane *p*-Xylene	Gulf Oil Chemicals Co.	N	–
None	Acetone, Phenol	Benzene Propylene	Allied Chemical	U	(17)
Cracking	Acetylene, Ethylene	Hydrocarbons (C_1-C_8)	Union Carbide	3	(10)

None	Acrolein, Acrylic acid, Acetaldehyde	Propylene	Standard Oil of Ohio	4	(2)
Esterification	Acrylic acid, Acrylic esters	Propylene Alcohols	BP Chemicals International Ltd.	1	(9)
			Mitsubishi Petrochemicals Ltd.	1	–
			Nippon Shokubai Kagaku Kogyo Co. Ltd.	1	–
			Toyo Soda Manufacturing Co., Inc.	N	–
None	Adipic acid	Cyclohexane	Gulf Oil Chemicals Co.	N	(6)
None	Adipic acid	Cyclohexyl alcohol	Veba-Chemie	2	–
			Zimmer/El Paso	1	–
None	Anthraquinone	Anthracene	Nippon Steel Chemical	U	(1)
			Toms River Chemical	U	–
None	Benzoic acid	Toluene	SNIA Viscosa	3	(5)
None	*tert*-Butyl alcohol	Isobutane	–	–	(2)
Beckmann Rearrangement Oximation	Caprolactam	Cyclohexane	Stamicarbon	13	(3)
			Inventa AG	6	–
Acid Rearrangement Hydrogenation	Caprolactam	Toluene	SNIA Viscosa	2	–
None	Cyclohexanone Cyclohexyl alcohol	Cyclohexane	Institut Francais du Petrole	1	(6)
			Scientific Design	8	–
			Stamicarbon	12	–
			Zimmer/El Paso	1	–
Esterification	Dimethyl terephthalate	Methanol *p*-Xylene	Dynamit Nobel	25	(7)
			Katzchmann/Simon Carves	14	–
			C. W. Witten	14	–

Table 3.2, continued

Other Required Processes	Product	Feedstock	Owner of Process	No. of Licensed Commercial Plants[a]	Total No. of U.S. Commercial Plants[b]
Esterification	Ethyl acrylate	Ethyl alcohol Propylene	–	–	(5)
None	Ethylene oxide	Ethylene	Shell	25	(15)
			Scientific Design	5	–
Hydrolysis	Ethylene oxide/Ethylene glycol	Ethylene	Hüls	4	–
			Japan Catalytic Chemical Ind. Co.	6	–
			Nippon Shokubai Kogako Kogyo Co. Ltd.	3	–
			Scientific Design (Ethylene oxide only)	49	–
			Shell	30	–
			SNAM Progetti S.p.A.	3	–
			Union Carbide	2	–
None[g]	Formaldehyde	Methanol	BASF AG	5	(54)
			Borden	U	–
			HIAG	1	–
			ICI	5	–
			Meissner	10	–
			Monsanto	7	–
None[h]	Formaldehyde	Methanol	Akita	1	–
			Chemico	U	–

			C'dF Chimie-IFP-Societe Chimique des Charbonnages	N	–
			Lummus	3	–
			Montedison S.p.A.	4	–
			Reichhold	49	–
None[h]	Formaldehyde	Dimethyl ether	Akita	1	–
None	Formic acid	Light hydrocarbons	BP Chemicals International Ltd.	3	(2)
Isomerization	Fumaric acid (*trans*-1,2-ethylene dicarboxylic acid)	Benzene	Scientific Design	5	(5)
None	Glycerine	Propylene	–	–	(4)
None	Isophthalic acid	*m*-Xylene	Standard Oil Co. (Indiana)	4	(1)
None	Maleic anhydride	Butadiene (+ other C_4 hydrocarbons)	BASF AG	2	(10)
None	Maleic anhydride	Benzene	Alsuisse-UCB	3	–
			Rhone Progil	1	–
			Scientific Design	26	–
			Veba Chemie AG/Bayer	U	–
None	Maleic anhydride	Butene-1, Butene-2, Butadiene (if present)	Bayer-Lurgi Corp.	1	–
Esterification	Methyl methacrylate	Isobutylene, Methanol	Gulf Oil Chemicals Co.	N	(4)
Ozonolysis	Pelargonic acid Caproic acid Azeliac acid	Oils (tall, red, soybean)	Welsbach Corp.	U	–
Ozonolysis	Pelargonic acid Undecanoic acid Tridecanoic acid	α-Olefins	Welsbach Corp.	N	–

Table 3.2, continued

Other Required Processes	Product	Feedstock	Owner of Process	No. of Licensed Commercial Plants[a]	Total No. of U.S. Commercial Plants[b]
Acid Cleavage	Phenol Acetone	Cumene	BP Chemicals International Ltd. and Hercules Inc.	22	(17)
			Rhone-Poulenc, S.A.	5	–
			UOP Process Division	3	–
Acid Cleavage Alkylation	Phenol Acetone	Benzene, Propylene	Allied Chemical	U	–
Dehydrogenation	Phenol, Hydrogen	Cyclohexane	Scientific Design	1	–
			Institut Francais du Petrole	N	–
None	Phthalic anhydride	Naphthalene	Nippon Steel Chemical	U	(11)
			Scientific Design	8	–
			Sherwin Williams/Badger	14	–
			United Chemicals & Coke, Ltd.	U	–
			Von Heyden	U	–
Alkylation	Phthalic anhydride	*o*-Xylene	Alusuisse	3	–
			BASF AG	20	–
			Rhone-Poulenc S.A.	7	–
			Scientific Design	4	–
			Von Heyden/Wacker	65	–
Alkylation	Pyromellitic dianhydride (1,2,4,5-benzenetetracarboxylic-1,2,4,5-dianhydride)	Pseudocumene	Gelsenberg-Chemie GmbH	N	–
Ozonolysis	Suberic acid Dodecanoic acid	Cyclic olefins	Welsbach Corp.	N	–

None	Terephthalic acid	*p*-Xylene	Eastman Kodak Co.-Tennessee Eastman Div.	1	(3)
			Institut Francais du Petrole	N	–
			Lummus	U	–
			Standard Oil Co. (Indiana)	26	–
			Toray Ind. Inc.	2	–
			Uni-Hüls	1	–
Condensation	Tetrahydrofuran, 2,3,4,5-tetracarboxylic dianhydride	Furan Maleic anhydride	Gelsenberg Chemie GmbH	N	(2)
Condensation	Tetramethylthiuram disulfide (thiuram) [*bis*(dimethylthiocarbamoyl)disulfide]	Ammonia, Carbon disulfide, Dimethylamine, Hydrogen peroxide	UCB	1	(5)
None	Trimellitic anhydride (1,2,4-benzene-tricarboxylic acid, 1,2-anhydride)	Psuedocumene	Standard Oil Co. (Indiana)	1	(1)

Major Unit Processes–Oxyhalogenation

None	Ethylene dichloride	Ethylene	BF Goodrich	17	–
			Monsanto	5	–
			Rhone Progil	2	–
			Stauffer Chemical Co.	8	–
Halogenation Cracking (catalytic)	Perchloroethylene Trichloroethylene	Any C_2 chlorocarbon mixture	PPG Industries	U	–
Hydrolysis	Phenol	Benzene	Hooker	3	–
		Hydrogen chloride	Union Carbide	1	–
Dehydro-halogenation Halogenation	Vinyl chloride monomer (VCM)	Ethane Chlorine	Transcat/Lummus Co.	U	–

Table 3.2, continued

Other Required Processes	Product	Feedstock	Owner of Process	No. of Licensed Commercial Plants[a]	Total No. of U.S. Commercial Plants[b]
Dehydro-halogenation	Vinyl chloride monomer (VCM)	Ethylene	B. F. Goodrich	31	(14)
Halogenation		Chlorine	Montedison S.p.A.	1	–
			Monsanto/Scientific Design	6	–
			P.P.G. Industries	U	–
			Rhone Progil	2	–
			Stauffer Chemical Co.	22	–
			Union Carbide	1	–
Dehydro-halogenation	Vinyl chloride monomer (VCM)	Naphtha	Kureha-Chiyoda	3	–
Halogenation		Chlorine	Nippon Zeon Co. Ltd.	1	–
		Major Unit Processes–Phosgenation			
None	Diphenylmethane-4,4′-diisocyanate [methylenebis(4-phenyl isocyanate)] (MDI)	Aniline Formaldehyde Phosgene	Hodogaya Chemical Co.	U	(4)
Dehydro-halogenation	Polycarbonate resins	Bisphenol A	Idemitsu Kosan	1	(3)
Polymerization		Phosgene			
Hydrogenation	Toluene diisocyanates (TDI)	Phosgene	Allied Chemical	U	(10)
Nitration	(80/20 2,4-2,6-TDI)	Toluene	FMC Corp.	N	–
			Nippon Soda-Nissan	U	–
			Sumitomo Chemical Co.	1	–

Major Unit Processes–Polymerization

None	Acrylic resins	Acrylonitrile	Zimmer AG	3	(122)
None	Acrylonitrile-butadiene-styrene resins (ABS resins)	Acrylonitrile	B. F. Goodrich	1	(14)
		Butadiene	Japan Synthetic Rubber Co.	U	–
		Styrene	Toray Ind. Inc.	2	–
			Toyo Koatsu	U	–
			Uniroyal	U	–
None	Epoxy resins	Bisphenol A Epichlorohydrin	Ciba-Geigy Corp.	U	(94)
None	Ethylene-propylene copolymers (EPM)	Ethylene	Montedison S.p.A./ B. F. Goodrich	2	(4)
		Propylene	Uniroyal	U	–
None	Ethylene-propylene terpolymer (EPT)	Ethylene	Montedison S.p.A./ B. F. Goodrich	2	(5)
		Propylene	Solvay & Cie	U	–
			Stamciarbon	2	–
None	Ethylene-vinyl acetate copolymer resins	Ethylene, Vinyl acetate	Wacker-Chemie	1	(15)
None	Polyamide resins (Nylon 6)	Caprolactam	Beaunit Corp.	U	–
			Inventa AG	20	(9)
			Lurgi	8	–
			Toray Ind., Inc.	U	–
			Ube Ind.	U	–
			Zimmer AG	109	–
None	Polyamide resins (Nylon 66)	Adipic acid Hexamethylenediamine	Beaunit Corp.	U	(3)

Table 3.2, continued

Other Required Processes	Product	Feedstock	Owner of Process	No. of Licensed Commercial Plants[a]	Total No. of U.S. Commercial Plants[b]
None[i]	*cis*-1,4-Polybutadiene	Butadiene	Hüls	1	–
			Japan Synthetic Rubber	U	–
			Phillips Petroleum Co.	8	–
			Polysar International	U	–
			SNAM Progetti	N	–
None[j]	Polybutadiene	Butadiene	SNPA	1	(5)
			Uniroyal	U	–
None[i]	Polybutadiene	Butadiene	Phillips Petroleum Co.	7	–
None[j]	Polybutadiene-acrylonitrile (NBR)	Acrylonitrile	Polysar International	U	(9)
		Butadiene	Uniroyal	U	–
None[j]	Polybutenes	Butene-1/Butene-2	Hüls	1	(6)
			Petro-Tex Chemical Corp.	N	–
Dehydro-halogenation	Polycarbonate resins	Bisphenol A	Idemitsu Kosan	1	(3)
Phosgenation		Phosgene			
None	Polychloroprene (neoprene)	Chloroprene	BP Chemicals International Ltd.	3	(3)
			Denki Kagaku	1	–
None	Polyester resins (saturated)	Glycols	BP Chemicals International Ltd.	U	(24)
		Polybasic acids			
		Styrene			
None	Polyester resins (unsaturated)	Glycols	BP Chemicals International Ltd.	U	(86)

		Styrene			
		Unsaturated dibasic acids			
None	Polyether glycols	Ethylene oxide	Lankro	4	(43)
		Propylene oxide			
None	Polyether glycols	Alcohols	Lankro	4	–
		Ethylene oxide	Mitsui Chemical	U	–
		Propylene oxide	Nippon Soda	1	–
			Rhone Progil	U	–
None	Polyethylene (low density)	Ethylene	ANIC	3	(20)
			ARCO	3	–
			ATO Chemie	14	–
			Dart	U	–
			Ethylene Plastique	7	–
			Gulf Oil Chemicals Co.	4	–
			Imperial Chemical Ind.	60	–
			Phillips Petroleum Co.	17	–
			SNAM Progetti	2	–
			Stamicarbon	10	–
			Sumitomo Chemical Co. Ltd.	2	–
			V/O Licensintorg	3	–
None	Polyethylene (high density)	Ethylene	UDHE GmbH	U	(14)
			Hoechst	U	–
			Hüls-Veba Chemie	4	–
			Imperial Chemical Industries	U	–
			Mitsubishi Chemical Industries	1	–
			Montedison S.p.A.	1	–
			Naphthachemie	1	–
			Phillips Petroleum Co.	17	–
			SNAM Progetti	1	–
			Solvay	5	–

Table 3.2, continued

Other Required Processes	Product	Feedstock	Owner of Process	No. of Licensed Commercial Plants[a]	Total No. of U.S. Commercial Plants[b]
			Stamicarbon	3	–
			Standard Oil	3	–
			Union Carbide	6	–
			Wacker-Chemie	2	–
Esterification	Polyethylene terephthalate	Ethylene glycol	Hoechst	U	(21)
		Terephthalic acid	ICI	U	–
			Inventa-Ems	12	–
			Toray Ind., Inc.	U	–
			Zimmer AG	36	–
Condensation	Polyethylene terephthalate	Dimethyl terephthalate	Beaunit Corp.	U	(21)
		Ethylene glycol	Hoechst	U	–
			ICI	11	–
			Inventa-Ems	4	–
			Zimmer AG	26	–
None	Polyisobutylene	Isobutene	Chevron Research	U	–
		Butenes	Cosden Oil & Chem.	5	–
Pyrolysis	Polyisocyanate	Organic dichlorides Sodium cyanate	Marathon Oil Co.	N	(1)
None[i]	*cis*-Polyisoprene (IR)	Isoprene	B. F. Goodrich	3	(2)
			SNAM Progetti	1	–

None	Polypropylene	Propylene	Friedrich UHDE GmbH	U	(12)
			Hercules	3	–
			Hoechst	U	–
			Hüls-Veba Chemie AG	2	–
			ICI	U	–
			Mitsu-Petrochemical Ind.-	15	–
			Montedison, S.p.A.	3	–
			Phillips Petroleum Co.	5	–
			Standard Oil Co. (Indiana)		
None	Polystyrene	Styrene	ATO	1	(36)
			Bakol-Scientific Design	1	–
			BP Chemicals International Ltd.	1	–
			Cosden Oil & Chemical Co.	6	–
			Hüls	1	–
			Polysar International	1	–
			Rhone Progil	1	–
			Standard Oil Co. (Indiana)	2	–
			Toyo Koatsu	U	–
			Union Carbide	11	–
None[k]	Polystyrene (high impact–rubber modified)	Polybutadiene	Cosden Technology, Inc.	U	(3)
		Styrene	Hoechst	1	–
None	Polyvinyl acetate	Vinyl acetate	Borden	U	(98)
			Wacker-Chemie	4	–
None	Polyvinyl alcohol resins	Vinyl alcohol	Aicello	U	(4)
None	Polyvinyl chloride resins	Vinyl chloride	ATO Chemie	1	(37)
			B. F. Goodrich	14	–
			Borden	U	–
			Diamond Shamrock	U	–
			Hoechst	5	–
			Houdry Div., Air Prod. & Chem.	2	–

Table 3.2, continued

Other Required Processes	Product	Feedstock	Owner of Process	No. of Licensed Commercial Plants[a]	Total No. of U.S. Commercial Plants[b]
			Hüls	5	–
			Kureha Chemical Co.	3	–
			Montedison S.p.A.	7	–
			Rhone Poulenc	39	–
			Scientific Design	13	–
			Societé National des Petroles D'Aguitaine	1	–
			Solvay-ICI	1	–
			Stauffer Chemical Co.	2	–
			Sumitomo Chemical Co.	7	–
			Union Carbide	U	–
			Uniroyal	U	–
			Wacker-Chemie	6	–
None	Polyvinyl chloride-acetate copolymer	Vinyl acetate Vinyl chloride	Scientific Design	12	(19)
None	Polyvinyl chloride-vinylidene chloride copolymer resins	Vinyl chloride Vinylidene chloride	Solvay	1	(9)
None	Propylene tetramer (dodecene, nonlinear)	Propylene	Chevron Research	6	(7)
			Phillips Petroleum Co.	U	–
			Texaco Development Corp.	U	–
			UOP Process Division	U	–
None[j]	SBR (polybutadiene-styrene)	Butadiene Styrene	American Synthetic Rubber Corp.	U	(10)
			B. F. Goodrich	1	–

			Hüls	1	–
			Phillips Petroleum Co.	U	–
			Polysar International	U	–
			Uniroyal	U	–
None[i]	SBR (polybutadiene-styrene)	Butadiene Styrene	Phillips Petroleum Co. Uniroyal	7 N	– –
None	Urea-formaldehyde resins	Biuret Formaldehyde Urea	Nipak	N	(126)

Major Unit Processes–Pyrolysis

Condensation	Acetic anhydride	Acetic acid	Wacker-Chemie	12+	(7)
None[l]	Acetylene	Hydrocarbons (C_1-C_8)	Hüls	1	(10)
None	*n*-Butyl acrylate	Acetic acid *n*-Butyl alcohol	Wacker-Chemie	N	(5)
None	Cyanuric acid (2,4,6-trihydroxy-1,3,5-triazine)	Urea	Gulf Oil Chemicals Co.	N	(2)
Halogenation	Cyanuric acid, Sodium dichloroisocyanurate, Trichloroisocyanuric acid	Caustic soda Chlorine Urea	Shikoku Kasei	3	–
Esterification	Ethyl acetoacetate	Acetic acid Ethanol	Wacker-Chemie	U	(2)
None[m]	Ethylene, Propylene, Hydrogen, Pyrolysis gasoline	Butanes, or Propane, or Ethane, or Naphtha or Gas Oil	C. F. Braun	22	(36)
			Fluor Corp. (ethane & gas oil only)	U	–
			Foster Wheeler	U	–
			Institut Francais du Petrole	U	–
			M. W. Kellog Co.	40	–

Table 3.2, continued

Other Required Processes	Product	Feedstock	Owner of Process	No. of Licensed Commercial Plants[a]	Total No. of U.S. Commercial Plants[b]
			Linde AG	100	–
			Lummus	85	–
			Monsanto	U	–
			Selas	U	–
			Stone & Webster	90	–
			UOP Process Div.	U	–
Halogenation	Hexachlorocyclopentadiene	Pentane	–	–	(3)
Condensation	Isoprene	Formaldehyde	Bayer	N	(6)
Hydro-halogenation		Hydrogen chloride Isobutylene	Marathon Oil	N	–
None	Ketene dimer (diketene)	Acetic acid	FMC Corp.	2	(2)
Condensation	Melamine	Dicyandiamide	Produits Azole	U	(3)
Condensation	Melamine	Urea	Montedison S.p.A.	3	–
			Nissan Chemical Industries	1	–
Esterification	Methyl acetoacetate	Acetic acid Isopropanol	Wacker-Chemie	U	(1)
None	Naphthalene	Coal tar Petroleum	–	–	(12)
Condensation	Oxalic acid	Sodium formate	Stauffer Chemical Co.	1	(7)
None	Phenothiazine	Diphenylamine Sulfur	Uniroyal	U	(1)

Polymerization	Polyisocyanate	Organic dichlorides Sodium cyanate	Marathon Oil Co.	N	(1)
None[m]	Propylene (see ethylene)	Propane, or Butane or Naphtha or Gas oils	–	–	(61)
		Major Unit Processes–Reforming (Steam)-Water Gas Reaction			
(High pressure)	Methanol	Naphtha	Chemical Construction Co.	U	(14)
			C & I, Girdler	U	–
			Haldor Topsoe	U	–
			Imperial Chemical Ind.	U	–
			M. W. Kellogg	U	–
			Montedison S.p.A.	U	–
			UKW	U	–
(High pressure)	Methanol	Natural gas	Borden Chem. Co.	U	–
			Chemical Construction Corp.	U	–
			C & I/Girdler	U	–
			Haldor Topsoe	U	–
			Imperial Chem. Inc.	U	–
			M. W. Kellogg	U	–
			Lummus	2	–
			Montedison S.p.A.	U	–
			Vulcan Cincinnati	13	–
(Low pressure)	Methanol	Naphtha	Imperial Chemical Ind.	20	–
			Lurgi	5	–
(Low pressure)	Methanol	Natural gas	Imperial Chemical Ind.	U	–
(Low pressure)	Methanol	Liquefied petroleum gas	Imperial Chemical Ind.	U	–
(Low pressure)	Methanol, Dimethyl ether	Naphtha, Carbon dioxide	Vulcan Cincinnati	13	–

Table 3.2, continued

Other Required Processes	Product	Feedstock	Owner of Process	No. of Licensed Commercial Plants[a]	Total No. of U.S. Commercial Plants[b]
	Major Unit Processes–Sulfonation				
Hydrolysis	Alcohols, mixed linear sulfated ammonium salt, sodium salt, triethanolamine salt, unspecified	Fatty alcohols, Sulfur trioxide	Chemithon	U	(9)
			MoDo Kemi AB	5	(2)
				–	(4)
				–	(2)
Hydrolysis	Alkylbenzene sulfonates	Alkylbenzenes, Sulfur trioxide	Chemithon	U	(10)
			Hüls	1	–
			MoDo Kemi AB	5	–
Hydrogenation Hydrolysis	*m*-Aminophenol	Nitrobenzene	Sumitomo Chem. Co.	1	(0)
Hydrolysis	*sec*-Butyl alcohol	Butylene	Scientific Design/ Maruzen Oil Co.	1	(4)
None	Ethoxylsulfates	Ethoxylates, Sulfur trioxide	Chemithon	U	–
			MoDo Kemi	5	–
Hydrolysis	Isopropyl alcohol	Propylene	BP Chemicals, Int., Ltd.	4	(5)
			Texaco Development Corp.	U	–
Esterification Hydrocyanation Hydrolysis	Methyl methacrylate monomer	Acetone Hydrogen cyanide Methanol	Montedison S.p.A.	1	(4)
			Sumitomo Chem. Co.	1	–
Hydrolysis	2-Naphthol (*beta*-naphthol)	Naphthalene Caustic soda	Mitsui Chem. Co.	U	(1)

Hydrolysis	Phenol	Benzene Caustic soda Sulfuric acid	Mitsui Chem. Co.	U	(1)

[a]U–At least one plant exists, but total is unknown.
[b]N–No commercial plants.
[c]Reductive alkylation.
[d]Catalytic dehydrogenation.
[e]Oxidative dehydrogenation.
[f]A direct catalytic hydrolysis process.
[g]Silver catalyst.
[h]Metal oxide catalyst.
[i]Solution polymerization.
[j]Emulsion polymerization.
[k]Suspension polymerization.
[l]Pyrolysis by electric arc.
[m]Steam pyrolysis.

Table 3.3 Minor Unit Processes–Summary

Process	Compounds	Entries to Table
1. Acid Cleavage	2	2
2. Acid Rearrangement	3	3
3. Amination by Reduction	2	2
4. Beckmann Rearrangement	1	2
5. Benzidine Rearrangement	1	1
6. Cannizzaro Reaction	1	1
7. Carboxylation	5	3
8. Chlorohydrination	3	4
9. Electrohydrodimerization	1	1
10. Epoxidation	4	3
11. Hydroacetylation	1	1
12. Hydrocyanation	2	2
13. Isomerization	3	5
14. Oximation	1	2
15. Oxyacetylation	1	1
16. Ozonolysis	7	3
TOTAL	38	36

Table 3.4 Minor Unit Processes

Processes	Product	Feedstock	Owner of Process	No. of Licensed Commercial Plants[a]	Total No. of U.S. Commercial Plants[b]
Acid Cleavage	Phenol	Cumene	BP Chemicals Int. Ltd. & Hercules	22	(17)
Oxidation	Acetone		Rhone Poulenc	5	–
			UOP Process Division	3	–
Alkylation	Phenol	Benzene	Allied Chemical	U	–
Hydrolysis	Acetone	Propylene			
Oxidation					
Acid Rearrangement	Caprolactam	Toluene, Ammonia	SNIA Viscosa	2	–
Hydrogenation					
Oxidation					
Acid Rearrangement	*m*-Dichlorobenzene (1,3-dichlorobenzene)	Dichlorobenzenes	–	–	(2)
Hydrogenation	*p*-Aminophenol	Nitrobenzene	–	–	(4)
Amination by Reduction	*o*-Aminophenol	Phenol	–	–	(2)
Nitration					
Benzidine Rearrangement	3,3′-Dichlorobenzidine dihydrochloride	1-Chloro-2-nitrobenzene	–	–	(2)
Beckmann Rearranagement	Caprolactam	Cyclohexane	Stamicarbon	13	(3)
Oxidation		Ammonia, Oleum	Inventa AG	6	–
Oximation					
Oximation	Caprolactam	Cyclohexanol/ cyclohexanone	Stamicarbon	4	–
		Ammonia, Oleum	Zimmer AG	4	–

Table 3.4, continued

Processes	Product	Feedstock	Owner of Process	No. of Licensed Commercial Plants[a]	Total No. of U.S. Commercial Plants[b]
Benzidine Rearrangement	3,3′-Dichlorobenzidine	1-Chloro-2-nitrobenzene	–	–	(2)
Amination by Reduction	dihydrochloride				
Cannizzaro Reaction	Pentaerythritol	Acetaldehyde	Montedison S.p.A.	1	(4)
Condensation		Formaldehyde	Meissner	U	–
Carboxylation	3-Hydroxy-2-naphthoic acid	*beta*-Naphthol, Carbon dioxide	Sumitomo Chem. Co.	2	(4)
Esterification	*p*-Oxybenzoic acid and *p*-Oxybenzoic bytyrate	Butyl alcohol Carbon dioxide, Phenol	Mitsui Chemical	U	–
Carboxylation	Salicylic acid	Phenol, Carbon dioxide, Caustic soda	Mitsui Chemical		
Carboxylation	Sodium *p*-aminosalicylate	*m*-Aminophenol, Caustic soda	Sumitomo Chem. Co.	1	(1)
Chlorohydrination	Epichlorohydrin	Allyl chloride	Union Carbide	1	(3)
Hydrolysis		Hypochlorous acid			
Halogenation, Hydrolysis	Glycerine (glycerol)	Propylene→allyl chloride→ epichlorohydrin	–	–	(2)
Chlorohydrination	Propylene oxide	Propylene	Lummus	N	(0)
Halogenation		*tert*-Butyl alcohol (recycled)			
Hydrolysis		Chlorine			
Hydrolysis	Propylene oxide	Propylene	BASF AG	1	(6)
		Chlorine solution	Hüls/Bayer	2	–
		Milk of lime	Union Carbide	>1	–

Electrohydrodimerization	Adiponitrile	Acrylonitrile	Asahi Chem. Ind. Co. Ltd.	1	–
			UCB S.A.	U	–
Epoxidation	Epoxidized polybutadiene	Acetic acid			
	(cured epoxy polybutadiene)	Hydrogen peroxide	FMC Corp.	U	–
		Polybutadiene			
Hydrolysis	Glycerine (glyercol)	Allyl alcohol	Daicel Ltd.	U	(1)
		Peracetic acid			
Epoxidation	Propylene oxide	Propylene	Daicel Ltd.	N	–
	Acetic acid	30% Peracetic acid			
		10-15% Acetic acid			
Hydroacetylation	Vinyl acetate	Acetylene	Borden-Blaw Knox	1	(2)
		Acetic acid	Scientific Design	1	–
			Wacker-Chemie	5	–
Hydrocyanation	Acrylonitrile	Acetylene	BF Goodrich	1	(5)
		Hydrogen cyanide	Zimmer AG	1	–
Esterification	Methyl methacrylate monomer	Acetone	Montedison S.p.A.	1	(4)
Hydrolysis		Hydrogen cyanide	Sumitomo Chem. Co.	1	–
Sulfonation		Methanol			
Isomerization	Fumaric acid	Maleic anhydride	Alusuisse	2	5
			Scientific Design	5	–
Oxidation	Fumaric acid	Benzene	Scientific Design	5	–
(catalytic)	*o*-Xylene	Aromatic mixtures	Atlantic-Engelhard Ind.	12	–
	p-Xylene		Toray Ind. Inc.-Isolene	2	–
			UOP Process Div.–Isomar	4	–
(catalytic with	*o*-Xylene	Xylenes, mixed	Mitsubishi Gas Chem. Co.	2	–
$HF\text{-}BF_3$)	*p*-Xylene				
(catalytic with	*o*-Xylene	Xylenes, mixed	ICI (Petrochemical Div.)	7	–
non noble metal)	*p*-Xylene				

Table 3.4, continued

Processes	Product	Feedstock	Owner of Process	No. of Licensed Commercial Plants[a]	Total No. of U.S. Commercial Plants[b]
Oximation	Caprolactam	Cyclohexane	Inventa AG	6	(3)
Beckmann Rearrangement		Ammonia	Stamicarbon	13	–
Oxidation		Oleum			
Oximation	Caprolactam	Cyclohexanol/ Cyclohexanone	Stamicarbon	4	–
Beckmann Rearrangement					
		Ammonia	Zimmer AG	4	–
		Oleum			
Oxyacetylation	Vinyl acetate	Ethylene	Bayer	8	(4)
		Acetic acid	National Distillers	2	–
		Oxygen			
Ozonolysis	Pelargonic acid	Oils (tall, red, soya bean)	Welsbach Corp.	U	–
Oxidation	Caproic acid	Ozone			
	Azeliac acid				
Oxidation	Pelargonic acid	α-Olefins	Welsbach Corp.	N	–
	Undecanoic acid	Ozone			
	Tridecanoic acid				
Oxidation	Suberic acid	Cyclic olefins	Welsbach Corp.	N	–
	Dodecanoic acid	Ozone			

[a]U–At least one plant exists, but total is unknown.
N–No commercial plants.
[b]Parentheses indicate total number of U.S. producing plants regardless of process.

Table 3.5 Index of Unit Processes and Feedstocks for Manufacture of 263 Commercial Organic Chemicals

	Product	Other Products	Processes	Feedstock
1.	Acetaldehyde	None	Dehydrogenation	Ethyl alcohol
	Acetaldehyde	None	Oxidation	Ethylene
2.	Acetic Acid	None	Alkylation Oxidation	*n*-Butenes
	Acetic Acid	None	Oxidation	Light naphtha
	Acetic Acid	Peracetic acid	Oxidation	Acetaldehyde
	Acetic Acid	Terphthalic acid	Oxidation	*n*-Butane, *p*-xylene
	Acetic Acid	Methanol	Hydrolysis	Methyl acetate
	Acetic Acid	None	Carbonylation (oxo)	Methanol, Carbon monoxide
3.	Acetic Anhydride	None	Condensation Pyrolysis	Acetic acid
4.	Acetone	None	Dehydrogenation	Isopropyl alcohol
	Acetone	Phenol	Acid cleavage, Alkylation Hydrolysis, Oxidation	Benzene, Propylene
	Acetone	Phenol	Acid cleavage, Oxidation	Cumene
5.	Acetylene	Ethylene	Oxidation Cracking (catalytic)	Hydrocarbons (C_1-C_8)
	Acetylene	None	Pyrolysis by electric arc	Hydrocarbons (C_1-C_8)
6.	Acrolein	Acetaldehyde Acrylic acid	Oxidation	Propylene
7.	Acrylamide	None	Hydrolysis	Acrylonitrile
8.	Acrylic Acid and Acrylate Esters	None	Esterification Oxidation	Propylene, Alcohols
9.	Acrylic Resins	None	Polymerization	Acrylonitrile

Table 3.5, continued

	Product	Other Products	Processes	Feedstock
10.	Acrylonitrile	None	Ammoxidation	Propylene
	Acrylonitrile	None	Hydrocyanation	Acetylene Hydrogen cyanide
11.	Acrylonitrile-Butadiene Styrene Resins (ABS resins)	None	Polymerization	Acrylonitrile, Butadiene, Styrene
12.	Adipic Acid	None	Oxidation	Cyclohexane
	Adipic Acid	None	Oxidation	Cyclohexyl alcohol
13.	Adiponitrile	None	Ammoxidation	Adipic acid
	Adiponitrile	Hexamethylene diamine	Halogenation Hydrogenation	Butadiene
	Adiponitrile	None	Electrohydrodimerization	Acrylonitrile
14.	Alcohols (C_7-C_{13})	None	Carbonylation (oxo)	Methanol, Carbon monoxide
15.	Alcohols, Mixed, Linear			
	Sulfated, Ammonium Salt	None	Sulfonation	Fatty alcohols, SO_3
	-, Sodium Salt	None	Sulfonation	Fatty alcohols, SO_3
	-, Triethanolamine Salt	None	Sulfonation	Fatty alcohols, SO_3
	-, Unspecified	None	Sulfonation	Fatty alcohols, SO_3
16.	Alkyl Benzenes (branched)	None	Alkylation	Benzene Propylene tetramer
	Alkyl Benzenes (linear)	None	Alkylation	Benzene, linear olefins
	Alkyl Benzenes (linear)	None	Alkylation Dehydrogenation	Benzene, linear parafins
17.	Alkylbenzene Sulfonates	None	Hydrolysis Sulfonation	Alkyl benzenes, Sulfur trioxide

18.	Allyl Chloride	None	Halogenation	Propylene
19.	*m*-Aminophenol	None	Hydrogenation Hydrolysis Sulfonation	Nitrobenzene
20.	*o*-Aminophenol	None	Amination by reduction Nitration	Phenol
21.	*p*-Aminophenol	None	Acid Rearrangement Hydrogenation	Nitrobenzene
22.	Aniline	None	Hydrogenation	Nitrobenzene
	Aniline	None	Ammination by ammonolysis	Phenol
23.	Anthraquinone	None	Oxidation	Anthracene
24.	Arsanilic Acid	None	Condensation	Aniline, Arsenic acid
25.	Azeliac Acid	Caproic acid Pelargonic acid	Ozonolysis Oxidation	Oils (tall, red, soya bean)
26.	Benzene	Xylenes	Alkylation	Toluene
	Benzene	None	Hydrodealkylation	Coke oven light oil
	Benzene	None	Hydrodealkylation	Toluene
	Benzene	None	Hydrodealkylation	Xylenes (mixed)
	Benzene	Light hydrocarbons	Hydrodealkylation	Aromatic mixtures
27.	Benzenesulfonamide	None	Amination by ammonolysis	Benzenesulfonyl chloride
28.	Benzenesulfonyl Chloride	None	Condensation	Benzene, Chlorosulfonic acid
29.	Benzoic Acid	None	Oxidation	Toluene
30.	Benzonitrile	None	Ammoxidation	Toluene
31.	Benzyl Chloride	None	Halogenation	Toluene
32.	Biphenyl	None	Condensation Dehydrogenation	Benzene

Table 3.5, continued

	Product	Other Products	Processes	Feedstock
33.	Bisphenol A	None	Condensation	Acetone, Phenol
34.	1,3-*bis*(aminomethyl)cyclohexane	None	Hydrogenation	Isophthalonitrile
35.	Bromoform	None	Halogenation	Acetone, Bromine
36.	Butadiene	None	Dehydrogenation (catalytic)	*n*-Butane
	Butadiene	None	Dehydrogenation (catalytic)	Butene-1, Butene-2
	Butadiene	None	Dehydrogenation (oxidative)	Butene-1, Butene-2
37.	*n*-Butyl Acetate	None	Esterification	Acetic acid, *n*-Butyl alcohol
38.	*n*-Butyl Acrylate	None	Pyrolysis	Acetic acid, *n*-Butyl alcohol
39.	*n*-Butyl Alcohol	*n*-Butyraldehyde	Carbonylation (oxo) Hydrogenation	Propylene, Carbon monoxide
	n-Butyl Alcohol	Crotonaldehyde *n*-Butyraldehyde	Condensation Hydrogenation	Acetaldehyde
40.	*sec.*-Butyl Alcohol	None	Hydrolysis Sulfonation	Butylene
	sec.-Butyl Alcohol	None	Hydrolysis	Butene-1, Butene-2
41.	*tert.*-Butyl Alcohol	None	Oxidation	Isobutane
42.	*n*-Butyl Benzyl Phthalate	None	Esterification	Benzyl alcohol, *n*-Butyl alcohol Phthalic anhydride
43.	*p-tert*-Butyl Phenol	None	Alkylation	Isobutene, Phenol
44.	*n*-Butyraldehyde	*n*-Butyl alcohol Crotonaldehyde	Condensation Hydrogenation	Acetaldehyde
	n-Butyraldehyde	*n*-Butyl alcohol		Propylene, Carbon monoxide

45	Caproic Acid	Azeliac acid Pelargonic acid	Oxidation Ozonolysis	Oils (tall, red, soya bean)
46.	Caprolactam	None	Acid rearrangement Hydrogenation Oxidation	Toluene, Ammonia
	Caprolactam	None	Beckmann rearrangement Oxidation Oximation	Cyclohexane, Ammonia Oleum
	Caprolactam	None	Beckmann rearrangement Oximation	Ammonia, Oleum Cyclohexanol, Cyclohexanone
47.	Carbon Tetrachloride Carbon Tetrachloride Carbon Tetrachloride	None Perchloroethylene Perchloroethylene	Halogenation Halogenation Halogenation	Carbon disulfide Methane, Propane Propane/Propylene
48.	Choline Chloride	None	Amination by ammonolysis Condensation Hydrohalogenation	Ethylene oxide Trimethyl amine
49.	Chloral	None	Halogenation	Acetaldehyde
50.	Chloroacetic Acid	None	Halogenation	Acetic acid
51.	Chlorobenzene (mono)	None	Halogenation	Benzene
52.	*p*-Chlorobenzene Sulfonamide	None	Amination by ammonolysis	*p*-Chlorobenzene sulfonyl chloride
53.	*p*-Chloro-*m*-cresol	None	Halogenation	*m*-Cresol
54.	Chloroform Chloroform	Chloromethanes Methyl chloride Methylene chloride	Halogenation Halogenation	Methane Methanol
55.	Chloronapthalene	None	Halogenation	β-Naphthol
56.	*o*-Chlorophenol	*p*-Chlorophenol	Halogenation	Phenol
57.	Chloropicrin	None	Halogenation	Picric acid

Table 3.5, continued

	Product	Other Products	Processes	Feedstock
58.	Chloroprene	None	Halogenation	Butadiene
59.	Crotonaldehyde	*n*-Butyl alcohol *n*-Butraldehyde	Condensation Hydrogenation	Acetaldehyde
60.	Cumene	None	Alkylation	Benzene Propylene
61.	Cyanuric Acid	Sodium dichloroisocyanurate, Trichloroisocyanuric acid	Halogenation Pyrolysis	Caustic soda Chlorine Urea
	Cyanuric Acid	None	Pyrolysis	Urea
62.	Cyclohexane	None	Hydrogenation	Benzene
63.	Cyclohexanone Cyclohexanone	None Cyclohexyl alcohol	Dehydrogenation Oxidation	Cyclohexane-Cyclohexanol mixtures Cyclohexane
64.	Cyclohexyl alcohol Cyclohexyl alcohol	None Cyclohexanone	Hydrogenation Oxidation	Phenol Cyclohexane
65.	Cyclohexylamine Cyclohexylamine	None None	Hydrogenation Hydrogenation	Aniline Nitrobenzene
66.	Di-*n*-butyl phthalate	None	Esterification	*n*-Butyl alcohol, Phthalic anhydride
67.	*m*-Dichlorobenzene *m*-Dichlorobenzene	None None	Acid rearrangement Halogenation	Dichlorobenzenes Benzene
68.	*o*-Dichlorobenzene	*p*-Dichlorobenzene	Halogenation	Benzene
69.	3,3′-Dichlorobenzidine dihydrochloride	None	Benzidine rearrangement, Amination by reduction	1-Chloro-2-nitrobenzene

70.	Dichlorodiphenyl- trichloroethane (DDT)	None	Condensation Halogenation	Acetaldehyde Monochlorobenzene
71.	Dichlorophenol	None	Halogenation	4-Chlorophenol, Phenol
72.	2,4-Dichlorophenoxy acetic acid (2,4-D)	None	Condensation Halogenation	Monochloroacetic acid Phenol
73.	2-(2,4-Dichlorophenoxy) propionic acid (2,4-DP)	None	Condensation Dehydrohalogenation	α-Chloropropionic acid 2,4-Dichlorophenol
74.	4,4′-Dichlorophenylsulfone	None	Condensation	Monochlorobenzene Sulfur trioxide
75.	1,2-Dichloropropane	None	Halogenation	Propylene
76.	1,3-Dichloropropane	None	Halogenation	Allyl chloride
77.	Diethyl Phthalate	None	Esterification	Ethyl alcohol, Phthalic anhydride
78.	Diheptyl Phthalate	None	Esterification	Heptyl alcohol, Phthalic anhydride
79.	Diisodecyl Phthalate	None	Esterification	Isodecyl alcohol, Phthalic anhydride
80.	Dimethylformamide	None	Amination by ammonolysis	Dimethylamine, Methyl formate
81.	Dimethyl Phthalate	None	Esterification	Methyl alcohol, Phthalic anhydride
82.	Dimethyl Terephthalate Dimethyl Terephthalate	None None	Esterification Esterification Oxidation	Methyl alcohol, Terephthalic acid Methyl alcohol *p*-Xylene
83.	4,6-Dinitro-*o*-cresol	None	Nitration	Cresol
84.	2,4-Dinitrophenol	None	Nitration	Phenol
85.	2,4-Dinitrotoluene 2,4-Dinitrotoluene	None 2,6-Dinitrotoluene	Nitration Nitration	Toluene Toluene
86.	Di-*n*-octyl Phthalate	None	Esterification	*n*-Octyl alcohol Phthalic anhydride

Table 3.5, continued

	Product	Other Products	Processes	Feedstock
87.	Dioctyl Phthalate (2-ethylhexyl phthalate)	None	Esterification	2-Ethylhexyl alcohol Phthalic anhydride
88.	Diphenylamine	None	Condensation	Aniline
89.	Diphenylguanidine	None	Condensation	Aniline Cyanic acid
90.	N,N′-Diphenylhydrazine (hydrazobenzene)	None	Condensation Hydrogenation	Nitrobenzene
91.	Diphenylmethane-4,4′-diisocyanate (MDI)	None	Phosgenation	Aniline, Formaldehyde, Phosgene
92.	Dodecanoic Acid	Suberic acid	Oxidation Ozonolysis	Cylcic olefins
93.	Dodecene, Nonlinear (propylene tetramer)	None	Polymerization	Propylene
94.	Epichlorohydrin	None	Chlorohydrination Hydrolysis	Allyl chloride Hypochlorous acid
95.	Epoxy Resins	None	Polymerization	Bisphenol A, Epichlorohydrin
96.	Ethanolamines	None	Amination by aminolysis	Ethylene oxide
97.	Ethoxysulfates	None	Sulfonation	Ethoxylates, Sulfur trioxide
98.	Ethyl Acetate Ethyl Acetate	None None	Condensation Esterification	Acetaldehyde Acetic acid, Ethyl alcohol
99.	Ethyl Acetoacetate	None	Esterification Pyrolysis	Acetic acid Ethyl alcohol

100.	Ethyl Acrylate	None	Carbonylation (oxo)	Acetylene, Ethyl alcohol Carbon monoxide
	Ethyl Acrylate	None	Esterification Oxidation	Ethyl alcohol, Propylene
101.	Ethyl Alcohol Ethyl Alcohol	None Ethyl ether	Hydrolysis Hydrolysis	Ethylene Ethylene
102.	Ethylamines	None	Amination by ammonolysis	Ethanol
103.	Ethyl Benzene	None	Alkylation	Benzene, Ethylene
104.	Ethyl Chloride	None	Hydrohalogenation	Ethylene
105.	Ethylene Ethylene	None Propylene Hydrogen Pyrolysis gasoline	Dehydration Pyrolysis	Ethyl alcohol Butanes, or Ethane, or Gas oil, or Naphtha, or Propane
106.	Ethylene Carbonate	None	Condensation	Ethylene oxide, Carbon dioxide
107.	Ethylenediamine Ethylenediamine	None None	Amination by ammonolysis Amination by ammonolysis	Ethylene dichloride Monoethanolamine
108.	Ethylene Dibromide	None	Halogenation	Ethylene
109.	Ethylene Dichloride Ethylene Dichloride	None None	Halogenation Oxyhalogenation	Ethylene Ethylene, Hydrogen chloride
110.	Ethylene Glycol	Ethylene oxide	Oxidation	Ethylene
111.	Ethylene Glycol Ethers	None	Condensation	Alkyl alcohols, Ethylene glycol
112.	Ethylene Glycol Monomethyl Ether	None	Condensation	Ethylene oxide, Ethanol
113.	Ethylene Oxide Ethylene Oxide	None Ethylene glycol	Oxidation Hydrolysis Oxidation	Ethylene Ethylene

Table 3.5, continued

	Product	Other Products	Processes	Feedstock
114.	Ethylene-Propylene Copolymers (EPM)	None	Polymerization	Ethylene, propylene
115.	Ethylene-Propylene Terpolymer (EPT)	None	Polymerization	Ethylene, Propylene
116.	Ethylene-Vinyl Acetate Copolymer Resins	None	Polymerization	Ethylene, Vinyl acetate
117.	Ethyl Ether	None	Condensation	Ethyl alcohol
118.	2-Ethyl Hexanol	None	Condensation Hydrogenation	Acetaldehyde Butyraldehyde
	2-Ethyl Hexanol	*n*-Butyl alcohol Isobutyraldehyde Isobutyl alcohol	Carbonylation (oxo) Hydrogenation	Propylene Carbon monoxide
119.	Ethyl Parathion	None	Condensation Halogenation	0,0-Dimethyl Phosphonothionochloridate Sodium nitrophenoxide
120.	Fluorocarbons	None	Halogenation	Carbon tetrachloride Hydrofluoric acid
	Fluorocarbons	None	Halogenation	Methane, Hydrofluoric acid
121.	Formic Acid	Sodium formate	Carbonylation (oxo) Hydrolysis	Carbon monoxide Sodium hydroxide
	Formic Acid	None	Carbonylation (oxo) Condensation Hydrolysis	Carbon monoxide (methanol recycled)
122.	Formaldehyde	None	Oxidation[a]	Methanol
	Formaldehyde	None	Oxidation[b]	Methanol

	Formaldehyde	None	Oxidation[b]	Dimethyl ether
123.	Formic Acid	None	Oxidation	Light hydrocarbons
124.	Fumaric Acid	None	Isomerization	Maleic anhydride
	Fumaric Acid	None	Isomerization Oxidation	Benzene
125.	Glycerine	None	Chlorohydrination Halogenation Hydrolysis	Propylene→allyl chloride→ epichlorohydrin
	Glycerine	None	Epoxidation Hydrolysis	Allyl alcohol Peracetic acid
	Glycerine	None	Oxidation	Propylene
126.	Heptenes	None	Condensation	Butylenes, Propylene
127.	Hexachlorobenzene	None	Halogenation	Benzene
128.	Hexachlorobutadiene	None	Halogenation	Butadiene
129.	Hexachlorocyclopentadiene	None	Pyrolysis	Pentane
130.	Hexachloroethane	None	Halogenation	Tetrachloroethane
131.	Hexamethylenediamine	None	Amination by ammonolysis	Adipic acid
	Hexamethylenediamine	None	Ammoxidation	Adipic acid
	Hexamethylenediamine	None	Hydrogenation	Adiponitrile
132.	Hexamethylenetetramine	None	Amination by ammonolysis Condensation	Ammonia Formaldehyde
90.	Hydrazobenzene[c] (N,N′-diphenylhydrazine)	None	Condensation Hydrogenation	Nitrobenzene
133.	Hydrogen Cyanide	None	Ammoxidation	Methane
134.	3-Hydroxy-2-Naphthoic Acid	None	Carboxylation	β-Napthol, Carbon dioxide
135.	α-Isoamylene	None	Dehydrogenation	Isopentane

Table 3.5, continued

	Product	Other Products	Processes	Feedstock
136.	Isobutyl Alcohol	*n*-Butyl alcohol 2-Ethyl hexanol Isobutyraldehyde	Carbonylation (oxo) Hydrogenation	Propylene Carbon Monoxide
	Isobutyl Alcohol	None	Carbonylation (oxo) Hydrogenation	Propylene Carbon Monoxide
137.	Isobutylene	None	Cracking (catalytic)	Isobutane
138.	Isobutyraldehyde	*n*-Butyl alcohol Isobutyl alcohol 2-Ethyl hexanol	Carbonylation (oxo) Hydrogenation	Propylene Carbon monoxide
139.	Isophthalic Acid	None	Oxidation	*m*-Xylene
140.	Isophthalonitrile	None	Ammoxidation	*m*-Xylene
141.	Isophorone	None	Condensation	Acetone
142.	Isoprene	None	Condensation Cracking (catalytic)	Propylene
	Isoprene	None	Condensation Dehydration Hydrogenation	Acetone Acetylene
	Isoprene	None	Condensation Cracking (catalytic)	Formaldehyde Isobutylene
	Isoprene	None	Condensation Hydrohalogenation Pyrolysis	Formaldehyde Isobutylene Hydrogen chloride
	Isoprene	None	Dehydrogenation	*tert*-Amylenes
	Isoprene	None	Cracking (catalytic)	Amylenes
	Isoprene	None	Dehydrogenation	Isopentane

143.	Isopropyl Acetate	None	Esterification	Acetic acid, Isopropyl alcohol
	Isopropyl Alcohol	None	Hydrolysis	Propylene
	Isopropyl Alcohol	None	Sulfonation Hydrolysis[d]	Propylene
144.	N-Isopropyl-N′-phenyl-*p*-phenylenediamine	None	Alkylation[e] Dehydrohalogenation Hydrogenation	Acetone Aniline *p*-Chloro-nitrobenzene
145.	Ketene Dimer	None	Pyrolysis	Acetic acid
146.	Lead Alkyls	None	Alkylation	Ethyl chloride, Alkyl chlorides
147.	Maleic Acid	None	Hydrolysis	Maleic anhydride
148.	Maleic Anhydride	None	Oxidation	Butadiene (plus other C_4 hydrocarbons)
	Maleic Anhydride	None	Oxidation	Benzene
	Maleic Anhydride	None	Oxidation	Butene-1, Butene-2, Butadiene (if present)
149.	Melamine	None	Condensation Pyrolysis	Dicyandiamide
	Melamine	None	Condensation Pyrolysis	Urea
150.	Mesityl Oxide	None	Condensation Dehydration	Acetone
151.	DL-Methionine	None	Condensation Hydrogenation	Acrolein Cyanic acid Methyl mercaptan
152.	Methyl Acetate	None	Esterification	Acetic acid Methyl alcohol
153	Methyl Acetoacetate	None	Esterification Pyrolysis	Acetic acid Isopropyl alcohol

Table 3.5, continued

	Product	Other Products	Processes	Feedstock
154.	Methyl Alcohol	None	Reforming (steam)[f]	Naphtha
	Methyl Alcohol	None	Reforming (steam)[f]	Natural gas
	Methyl Alcohol	None	Reforming (steam)[g]	Naphtha
	Methyl Alcohol	None	Reforming (steam)[g]	Natural gas
	Methyl Alcohol	None	Reforming (steam)[g]	Liquefied petroleum gas
	Methyl Alcohol	Methyl ether	Reforming (steam)[g]	Naphtha, Carbon dioxide
155.	Methylamines	None	Amination by ammonolysis	Methanol, Ammonia
156.	Methyl Bromide	None	Hydrohalogenation	Methanol
157.	2-Methyl-1-butanol (*tert*-amyl alcohol)	None	Condensation Hydrogenation	Acetone Acetylene
158.	2-Methyl-3-butyn-2-ol	None	Condensation	Acetone, Acetylene
159.	Methyl Chloride	None	Halogenation	Methanol
	Methyl Chloride	None	Halogenation	Methane
	Methyl Chloride	None	Hydrohalogenation	Methanol
160.	2-Methyl-4-chloro-phenoxyacetic Acid (MCPA)	None	Condensation Halogenation	*o*-Cresol Monochloroacetic acid
161.	2-(2-Methyl-4-chlorophenoxy) propionic Acid (MCPP)	None	Condensation Dehydrohalogenation	α-Chloropropionic acid, 4-Chloro-*o*-cresol
162.	Methylene Chloride	Chloroform Methyl chloride	Halogenation	Methyl alcohol
	Methylene Chloride	Chloroform Methyl chloride	Halogenation	Methane
	Methylene Chloride	Chloroform Methyl chloride	Halogenation	Methane and Methyl alcohol

163.	Methyl Ether	Methyl alcohol	Reforming (steam)	Naphtha
164.	Methyl Ethyl Ketone	None	Dehydrogenation Hydrolysis	Butene-1 Butene-2
	Methyl Ethyl Ketone	None	Dehydrogenation	*sec*.-Butyl alcohol
165.	2-Methyl-5-ethyl pyridine (MEP)	None	Amination by amminolysis Condensation	Acetaldehyde Ammonia
166.	Methyl Isobutyl Ketone	None	Condensation Dehydration Hydrogenation	Acetone Hydrogen
167.	Methyl Methacrylate Monomer	None	Esterification Hydrocyanation Hydrolysis Sulfonation	Acetone Hydrogen cyanide Methyl alcohol
	Methyl Methacrylate Monomer	None	Esterification Oxidation	Isobutylene Methyl alcohol
168.	Methyl Parathion	None	Condensation Halogenation	0,0-Dimethyl phosphorothionochloridate, Sodium *p*-nitrophenoxide
169.	4-Methyl-1-pentene	None	Condensation	Propylene
170.	Monochloroacetic Acid	None	Halogenation	Acetic acid
171.	Morpholine	None	Dehydration	Diethanolamine
172.	Naphthalene	None	Pyrolysis	Coal tar or Petroleum
	Naphthalene	None	Hydrodealkylation	Alkyl naphthalenes
173.	2-Naphthol	None	Hydrolysis Sulfonation	Naphthalene
174.	Nitrobenzene	None	Nitration	Benzene
175.	*o*-Nitrophenol	None	Nitration	Phenol

Table 3.5, continued

	Product	Other Products	Processes	Feedstock
176.	*p*-Nitrophenol	None	Nitration	Phenol
177.	*m*-Nitrotoluene	None	Nitration	Toluene
178.	*o*-Nitrotoluene	None	Nitration	Toluene
179.	*p*-Nitrotoluene	None	Nitration	Toluene
180.	*p*-Nonyl Phenol	None	Alkylation	Phenol Propylene trimer
181.	Oxalic Acid	None	Condensation Pyrolysis	Sodium formate
182.	*p*-Oxybenzoic Acid	*p*-Oxybenzoic butyrate	Carboxylation Esterification	Butyl alcohol Carbon dioxide, Phenol
183.	Pelargonic Acid	Azeliac acid Caproic acid	Oxidation Ozonolysis	Oils (tall, red soya bean)
	Pelargonic Acid	Undecanoic acid Tridecanoic acid	Oxidation Ozonolysis	Alpha-Olefins
184.	Pentachlorophenol Pentachlorophenol	None None	Halogenation Hydrolysis	Phenol Hexachlorobenzene
185.	Pentaerythritol	None	Cannizzaro Reaction Condensation	Acetaldehyde Formaldehyde
186.	Perchloroethylene	Trichloroethylene	Cracking (catalytic) Halogenation	Ethylene dichloride
	Perchloroethylene	Trichloroethylene	Cracking (catalytic) Halogenation	Acetylene
	Perchloroethylene	Trichloroethylene	Cracking (catalytic) Halogenation Oxyhalogenation	Any C_2 chlorocarbon mixture

187.	Phenol	Acetone	Acid cleavage Alkylation Hydrolysis Oxidation	Benzene, Propylene
	Phenol	Acetone	Acid cleavage Oxidation	Cumene
	Phenol	Hydrogen	Dehydrogenation Oxidation	Cyclohexane
	Phenol	None	Hydrolysis	Monochlorobenzene
	Phenol	None	Hydrolysis Sulfonation	Benzene, Sulfuric acid
	Phenol	None	Hydrolysis Oxyhalogenation	Benzene Hydrogen chloride
188.	Phenothiazine	None	Pyrolysis	Diphenylamine, Sulfur
189.	N-phenyl-2-naphthylamine	None	Condensation	Aniline, 2-naphthol
190.	*p*-Phenylphenol	None	Condensation Dehydrogenation	Benzene, Cyclohexanone
191.	Phosgene	None	Halogenation	Carbon monoxide, Chlorine
192.	Phthalic Anhydride	None	Oxidation	Naphthalene
	Phthalic Anhydride	None	Oxidation	*o*-Xylene
193.	Phthalonitrile	None	Ammoxidation	*o*-Xylene
194.	β-Picoline	Pyridine	Ammoxidation Condensation	Acetaldehyde Formaldehyde, Methanol
195.	Piperylene	None	Dehydrogenation	*n*-Pentene
196.	Polyamide Resins (nylon 6)	None	Polymerization	Caprolactam
197.	Polyamide Resins (nylon 66)	None	Polymerization	Adipic acid, Hexamethylene diamine

Table 3.5, continued

	Product	Other Products	Processes	Feedstock
198.	*cis*-1,4-Polybutadiene[h]	None	Polymerization	Butadiene
199.	Polybutadiene[i]	None	Polymerization	Butadiene
	Polybutadiene[h]	None	Polymerization	Butadiene
200.	Polybutadiene-Acrylonitrile (NBR)[i]	None	Polymerization	Acrylonitrile Butadiene
201.	Polybutadiene (epoxidized)	None	Epoxidation	Acetic acid, Hydrogen peroxide, Polybutadiene
202.	Polybutenes	None	Polymerization	Butene-1, Butene-2
203.	Polycarbonate Resins	None	Dehydrohalogenation Phosgenation Polymerization	Bisphenol-A, Phosgene
204.	Polychloroprene	None	Polymerization	Chloroprene
205.	Polyester Resins (saturated)	None	Polymerization	Glycols, Polybasic acids, Styrene
206.	Polyester Resins (unsaturated)	None	Polymerization	Glycols, Styrene, Unsaturated dibasic acids
207.	Polyether Glycols	None	Polymerization	Ethylene oxide, Propylene oxide
	Polyether Glycols	None	Polymerization	Alcohols, Ethylene oxide, Propylene oxide
208.	Polyethylene (low-density)	None	Polymerization	Ethylene
209.	Polyethylene (high-density)	None	Polymerization	Ethylene
210	Polyethylene Terephthalate	None	Condensation Polymerization	Dimethyl terephthalate, Ethylene glycol

No.	Product		Process	Raw materials
	Polyethylene Terephthalate	None	Esterification Polymerization	Ethylene glycol, Terephthalic acid
211.	Polyisobutylene	None	Polymerization	Isobutylene, Butenes
212.	Polyisocyanate	None	Polymerization Pyrolysis	Organic dichlorides Sodium cyanate
213.	*cis*-Polyisoprene (IR)[j]	None	Polymerization	Isoprene
214.	Polypropylene	None	Polymerization	Propylene
215.	Polystyrene	None	Polymerization	Styrene
216.	Polystyrene (high-impact rubber modified)[l]	None	Polymerization	Polybutadiene, Styrene
217.	Polyvinyl Acetate	None	Polymerization	Vinyl acetate
218.	Polyvinyl Alcohol Resins	None	Polymerization	Vinyl alcohol
219.	Polyvinyl Chloride Resins	None	Polymerization	Vinyl chloride
220.	Polyvinyl Chloride-Acetate Copolymer	ɔne	Polymerization	Vinyl acetate, Vinyl chloride
221.	Polyvinyl Chloride–Vinylidene Chloride Resins	›ne	Polymerization	Vinyl chloride Vinylidene chloride
222.	β-Propiolactone	None	Polymerization	Formaldehyde, Ketene
223.	Propylene Propylene	None None	Dehydrogenation Pyrolysis	Propane Butane, or Gas oils, or Naphtha, or Propane
224.	Propylene Carbonate	None	Condensation	Carbon dioxide Propylene oxide
225.	Propylene Glycol	None	Hydrolysis	Propylene oxide

Table 3.5, continued

	Product	Other Products	Processes	Feedstock
226.	Propylene Oxide	None	Chlorohydrination Halogenation Hydrolysis	Propylene Chlorine solution *tert*-Butyl alcohol (recycled)
	Propylene Oxide	Acetic acid	Epoxidation	Propylene 30% Peracetic acid 10-15% Acetic acid
	Propylene Oxide	None	Chlorhydrination Hydrolysis	Propylene Chlorine solution Milk of lime
(93.)	Propylene Tetramer[k] (Dodecene)	None	Polymerization	Propylene
227.	Pyridine	β-Picoline	Ammoxidation Condensation	Acetaldehyde Formaldehyde Methanol
228.	Pyromellitic Dianhydride	None	Alkylation Oxidation	1,2,4-Trimethylbenzene (pseudocumene)
229.	Salicylic Acid	None	Carboxylation	Phenol, Carbon dioxide, Caustic soda
230.	SBR (polybutadiene-styrene)[i]	None	Polymerization	Butadiene, Styrene
	SBR (polybutadiene-styrene)[h]	None	Polymerization	Butadiene, Styrene
231.	Sodium *p*-amino-salicylate	None	Carboxylation	*m*-Aminophenol, Carbon dioxide, Caustic soda
232.	Sodium Formate	None	Carbonylation (oxo)	Carbon monoxide, Sodium hydroxide

233.	Sorbitol	None	Hydrogenation	Corn sugar or Corn Syrup
234.	Styrene	None	Alkylation Dehydrogenation	Benzene, Ethylene
	Styrene	None	Dehydrogenation	Ethylbenzene
235.	Suberic Acid	Dodecanoic acid	Oxidation Ozonolysis	Cylcic olefins
236.	Terephthalic Acid	None	Oxidation	*p*-Xylene
237.	Terephthalonitrile	None	Ammoxidation	*p*-Xylene
238.	1,1,2,2-Tetrachloroethane	None	Halogenation	Ethane
239.	Tetrahydrofuran, 2,3,4,5-Tetracarboxylic dianhydride	None	Condensation Oxidation	Furan, Maleic anhydride
240.	Tetramethylthiuram Disulfide (Thiram)	None	Condensation Oxidation	Ammonia, Carbon disulfide, Dimethylamine, Hydrogen peroxide
241.	Toluene Diisocyanate (TDI) (80/20-2,4-2,6-TDI)	None	Hydrogenation Nitration Phosgenation	Phosgene Toluene
242.	Triacetin Polymer	None	Esterification	Acetic acid, Cellulose
243.	Tributyrin (glyceryl tributyrate)	None	Esterification	*n*-Butyric acid, Glycerol
244.	1,2,4-Trichlorobenzene	None	Halogenation	1,2-Dichlorobenzene
245.	1,1,1-Trichloroethane 1,1,1-Trichloroethane	None None	Halogenation Halogenation Hydrohalogenation	Ethylene Vinyl chloride
246.	1,1,2-Trichloroethane	None	Halogenation	Ethylene
247.	Trichloroethylene	Perchloroethylene	Cracking (catalytic) Halogenation	Ethylene dichloride

Table 3.5, continued

	Product	Other Products	Processes	Feedstock
	Trichloroethylene	Perchloroethylene	Cracking (catalytic) Halogenation	Acetylene
	Trichloroethylene	Perchloroethylene	Cracking (catalytic) Halogenation Oxyhalogenation	Any C_2 chlorocarbon mixture
248.	2,4,5-Trichlorophenol	None	Halogenation Hydrolysis	Benzene Methanol Sodium hydroxide
249.	2,4,5-Trichlorophenoxy-acetic Acid (2,4,5-T)	None	Condensation Halogenation	Acetic acid Trichlorophenol
250.	Tridecanoic Acid	Pelargonic acid Undecanoic acid	Oxidation Ozonolysis	α-Olefins
251.	Trimellitic Anhydride	None	Oxidation	Pseudocumene
252.	Undecanoic Acid	Pelargonic acid Tridecanoic acid	Oxidation Ozonolysis	α-Olefins
253.	Urea	None	Amination by ammonolysis Dehydration	Carbon dioxide
254.	Urea-Formaldehyde Resins	None	Polymerization	Biuret, Formaldehyde, Urea
255.	Vinyl Acetate	None	Hydroacetylation	Acetylene Acetic acid
	Vinyl Acetate	None	Oxyacetylation	Ethylene, Acetic acid, Oxygen
256.	Vinyl Chloride Monomer (VCM)	Hydrogen chloride	Dehydrohalogenation	Ethylene dichloride
	Vinyl Chloride Monomer (VCM)	None	Dehydrohalogenation Halogenation Oxyhalogenation	Ethylene Chlorine

No.	Product	Coproduct	Process	Raw materials
	Vinyl Chloride Monomer (VCM)	None	Dehydrohalogenation Halogenation Oxyhalogenation	Ethane Chlorine
	Vinyl Chloride Monomer (VCM)	None	Dehydrohalogenation Halogenation	Acetylene Chlorine
	Vinyl Chloride Monomer (VCM)	None	Dehydrohalogenation Halogenation Oxyhalogenation	Naphtha Chlorine
	Vinyl Chloride Monomer (VCM)	None	Hydrohalogenation	Acetylene, Chlorine
257.	Vinylidene Chloride	None	Dehydrohalogenation Halogenation	Vinyl chloride Chlorine
258.	*o*-Xylene[l]	*p*-Xylene	Isomerization	Aromatic mixtures
	o-Xylene[m]	*p*-Xylene	Isomerization	Xylenes, mixed
	o-Xylene[n]	*p*-Xylene	Isomerization	Xylenes, mixed
259.	Xylenes, mixed	--	Dehydrogenation	Naphtha
260.	*m*-Xylenediamine	None	Hydrogenation	Isophthalonitrile
261.	2,4-Xylenol	None	Alkylation	*p*-Cresol, Methyl chloride
262.	Zineb	None	Condensation	Carbon disulfide, Ethylenediamine, Zinc sulfate
263.	Ziram	None	Condensation	Carbon disulfide, Ethylenediamine, Zinc sulfate

[a]Silver catalyst.
[b]Metal oxide catalyst.
[c]Listed under N,N'-diphenylhydrazine as No. 90.
[d]Direct catalytic hydrolysis.
[e]Reductive alkylation.
[f]High pressure.
[g]Low pressure.
[h]Solution polymerization.
[i]Emulsion polymerization.
[j]Suspension polymerization.
[k]Listed under Dodecene as No. 93.
[l]Catalytic isomerization.
[m]Catalytic isomerization with $HF\text{-}BF_3$.
[n]Catalytic isomerization with nonnoble metal.

4

DEFINITIONS OF MAJOR UNIT PROCESSES

1. AKLYLATION

Alkylation is the introduction of an alkyl radical into an organic compound by substitution or addition. There are six general types of alkylation, depending on the linkage that is attached:

a. Substitution for Hydrogen Bound to Carbon

In this type of alkylation an alkyl group is substituted for a hydrogen atom, which is bonded to a carbon atom. An example is the formation of ethyl benzene by the reaction of ethylene with benzene in the presence of a catalyst, as shown below:

$$C_6H_6 + CH_2{=}CH_2 \xrightarrow[150\text{-}250^\circ C]{\text{catalyst}} C_6H_5{-}CH_2CH_3$$

b. Substitution for Hydrogen Attached to Nitrogen

In this type of alkylation an alkyl group is substituted for a hydrogen attached to a trivalent nitrogen atom. As an example, benzylethylaniline is made by the reaction of benzyl chloride with ethylaniline, as shown below:

$$C_6H_5{-}CH_2Cl + C_6H_5{-}\underset{H}{N}C_2H_5 \longrightarrow C_6H_5{-}CH_2\underset{C_6H_5}{N}C_2H_5 + HCl$$

c. Formation of Alkyl-Metallic Compounds

In this type of alkylation an alkyl group unites with a metal to form a carbon-to-metal bond. An example is the formation of tetraethyllead by the reaction of ethyl chloride with a lead sodium alloy at 65-75°C and 3-4 atmospheres pressure, as shown below:

$$4C_2H_5Cl + 4PbNa \rightarrow Pb(C_2H_5)_4 + 3Pb + 4NaCl$$

d. Substitution for Hydrogen in a Hydroxyl Group

In this type of alkylation an alkyl group is substituted for the hydrogen in the hydroxyl group of an alcohol or phenol. An example is the formation of anisole (methyl phenyl ether) by the reaction of methyl chloride with phenol in the presence of alkali.

$$CH_3Cl + C_6H_5OH + NaOH \rightarrow C_6H_5OCH_3 + NaCl + H_2O$$

e. Formation of Quaternary Ammonium Compounds

In this type of alkylation a quaternary ammonium compound is formed by the addition of an alkyl halide, an alkyl sulfate, or an alkyl sulfonate to a tertiary amine. An example is the reaction of dimethylaniline with methyl chloride to form phenyl trimethylammonium chloride, as shown below:

$$C_6H_5N(CH_3)_2 + CH_3Cl \rightarrow C_6H_5N^+(CH_3)_3Cl^-$$

f. Miscellaneous Alkylations

Other alkylations include the processes by which an alkyl group is bound to sulfur, as in mercaptans, and by which an alkyl group is bound to silicon, as in alkyl silanes. For example, ethyltrichlorosilane is formed by the reaction of ethylene and trichlorosilane in the presence of a peroxide catalyst, as shown below:

$$CH_2{=}CH_2 + SiHCl_3 \xrightarrow{\text{catalyst}} C_2H_5SiCl_3$$

Lauryl mercaptan (*n*-dodecyl mercaptan) is made by the reaction of lauryl chloride with sodium hydrosulfide, as shown below:

$$n\text{-}C_{12}H_{25}Cl + NaSH \rightarrow n\text{-}C_{12}H_{25}SH + HCl$$

2. AMINATION BY AMMONOLYSIS

Ammonolysis is the process of forming amines by using ammonia, or primary amines, or secondary amines as aminating agents. Hydroammonolysis, in which ammonia-hydrogen mixtures are used in the presence of a hydrogenation catalyst, is also included. This technique permits the direct preparation of amines from carbonyl compounds, which would form nitriles or aldimines with ammonia alone. Ammonolytic reactions may involve the following:

a. Double Decomposition

In this type of reaction the NH_3 molecule is split into $-NH_2$ and -H fragments. The $-NH_2$ fragment becomes part of the newly formed amine, while the -H fragment unites with the radical being substituted, such as -Cl. An example is the formation of ethylenediamine from ethylene dichloride, as shown below:

$$ClCH_2CH_2Cl + 2NH_3 \xrightarrow[110^\circ C]{} H_2NCH_2CH_2NH_2 + 2HCl$$

b. Dehydration

In the ammonolysis of alcohols ammonia may be considered a dehydrant, in that an amine and water are formed. An example is the formation of methylamine and water from the reaction of methanol and ammonia, as shown below:

$$CH_3OH + NH_3 \xrightarrow[450^\circ C]{\text{catalyst}} CH_3NH_2 + H_2O$$

c. Simple Addition

In this type of ammonolysis both the $-NH_2$ and -H fragments of ammonia are part of the newly formed amine. An example is the formation of ethanolamine by the reaction of ethylene oxide and ammonia, as shown below:

$$\underset{\diagdown O \diagup}{CH_2{-}CH_2} + NH_3 \rightarrow \underset{OH\quad\;\; NH_2}{CH_2{-}CH_2}$$

d. Multiple Activity

In this type of ammonolytic reaction the newly formed amine, or recycled amines, compete with ammonia as a reactant to produce secondary or tertiary amines. An example is the formation of monomethylamine, dimethylamine and trimethylamine in the ammonolysis of methanol, as shown below:

$$CH_3OH + NH_3 \longrightarrow CH_3NH_2 + HOH$$

$$CH_3OH + CH_3NH_2 \longrightarrow (CH_3)_2NH + HOH$$

$$CH_3OH + (CH_3)_2NH \longrightarrow (CH_3)_3N + HOH$$

3. AMMOXIDATION

Ammoxidation is a process in which nitriles are formed by the reaction of ammonia in the presence of air or oxygen, with olefins, organic acids or the alkyl group of alkylated aromatics.

An example is the manufacture of acrylonitrile by the ammoxidation of propylene in a fluidized bed reactor, operating at 400-510°C and 1-2 atm pressure. The equation for the reaction follows:

$$CH_2{=}CHCH_3 + NH_3 + 3/2O_2 \xrightarrow[400\text{-}510^\circ C]{\text{catalyst}} CH_2{=}CHCN + 3H_2O$$

4. CARBONYLATION (OXO)

Carbonylation, or the Oxo reaction, is the combination of an organic compound with carbon monoxide. It is a method of converting α-olefins to aldehydes and/or alcohols containing one additional carbon atom. The olefin in the liquid state is reacted with a mixture of hydrogen and carbon monoxide in the presence of a soluble cobalt catalyst, such as dicobalt octacarbonyl, to produce an aldehyde and some alcohol. All or part of the aldehyde may be hydrogenated over a nickel catalyst to form an alcohol.

An example is the BASF AG process for manufacture of *n*-butyraldehyde, in which propylene is reacted at 140-170°C and 270-300 atm pressure in the liquid phase with carbon monoxide and hydrogen in the presence of dissolved cobalt catalyst. The equation for the reaction follows:

$$CH_2{=}CHCH_3 + CO + H_2 \xrightarrow[160\text{-}175^\circ C]{\text{catalyst}} CH_3CH_2CH_2CHO$$

The Monsanto carbonylation process for acetic acid manufacture reacts liquid methanol with gaseous carbon monoxide at 175°C and 30 atm pressure in the presence of a soluble rhodium iodocarbonyl catalyst. The equation for the reaction is:

$$CH_3OH + CO \xrightarrow[175^\circ C]{\text{catalyst}} CH_3COOH$$

5. CONDENSATION

Condensation is a chemical reaction in which two or more molecules combine, usually with the separation of water or some other low-molecular-weight compound. Each of the reactants contributes a part of the separated compound. Condensation is used in making many different chemicals, as shown by the 55 listed in the Condensation Table.

A typical example of a condensation reaction is the combination of two moles of phenol with acetone to give bisphenol A and water. The reaction occurs with dry HCl as a catalyst at 50°C for 8-12 hours. The equation for the reaction follows:

$$2\,C_6H_5{-}OH + CH_3\overset{O}{\overset{\|}{C}}CH_3 \longrightarrow HO{-}C_6H_4{-}\underset{CH_3}{\overset{CH_3}{C}}{-}C_6H_4{-}OH + H_2O$$

6. CRACKING, CATALYTIC

Catalytic cracking is the thermal decomposition of an organic compound in the presence of a catalyst and in the absence of air. This book does not cover the use of catalytic cracking in the petroleum industry in which high-boiling virgin gas oils are converted to lower-boiling gasoline components.

Trichloroethylene is produced by the cracking of tetrachloroethane in the presence of the catalyst $BaCL_2$ on-carbon. The reaction takes place at 250-300°C. The formation of trichloroethylene is shown in the following equation:

$$CHCl_2CHCl_2 \xrightarrow[250\text{-}300^\circ C]{\text{catalyst}} \underset{Cl}{CH}{=}\underset{Cl}{C}Cl + HCl$$

7. DEHYDRATION

Dehydration is defined here as a decomposition reaction, in which a new compound and water are formed from a single molecule. This is chemical dehydration and does not include physical dehydration, in which a compound is dried by heat. Reactions in which two molecules condense with the elimination of water and the formation of a new compound are included in the unit process of condensation.

An example of dehydration is the formation of ethylene from ethanol, as shown in the following equation:

$$CH_3CH_2OH \xrightarrow[\text{heat}]{\text{catalyst}} CH_2{=}CH_2 + H_2O$$

8. DEHYDROGENATION

Dehydrogenation is the process by which a new chemical is formed by the removal of hydrogen from the feedstock compound. Aldehydes and ketones are prepared by the dehydrogenation of alcohols.

An example is the dehydrogenation of ethyl alcohol to acetaldehyde, as shown by the following equation:

$$CH_3CH_2OH \xrightarrow[250\text{-}350^\circ C]{\text{Cu catalyst}} CH_3\overset{\overset{\displaystyle O}{\|}}{C}H + H_2$$

Saturated hydrocarbons are dehydrogenated to olefins. As an example, *n*-butane is catalytically dehydrogenated to butadiene, as shown by the following equation:

$$CH_3CH_2CH_2CH_3 \xrightarrow[600\text{-}620^\circ C]{\text{chromia-alumina catalyst}} CH_2{=}CHCH{=}CH_2 + 2H_2$$

9. DEHYDROHALOGENATION

In the process of dehydrohalogenation, a hydrogen atom and a halogen atom are removed from one or more feedstocks to obtain a new chemical. In commercial operation, the halogen atom is usually chlorine.

The Transcat-Lummus process for the production of vinyl chloride monomer (VCM) uses dehydrochlorination as the last step in the process. Ethylene dichloride is dehydrohalogenated to vinyl chloride monomer, as shown in the following equation:

$$ClCH_2CH_2Cl \xrightarrow[480\text{-}510^\circ C]{\text{catalyst}} CH_2{=}CHCl + CHl$$

10. ESTERIFICATION

Esterification is the process by which an ester is formed. An ester is an organic compound derived from an organic acid and an alcohol by the exchange of the ionizable hydrogen atom of the acid for an organic radical. In transesterification, an ester reacts with an alcohol to form a different ester.

An example of esterification is the formation of ethyl acetate by the reaction of acetic acid and ethyl alcohol in the presence of a mineral acid catalyst, as shown by the following equation:

$$CH_3\overset{\overset{\displaystyle O}{\|}}{C}OH + C_2H_5OH \rightleftharpoons CH_3\overset{\overset{\displaystyle O}{\|}}{C}OC_2H_5 + H_2O$$

Dimethylterephthalate, used in the production of polyesters (polyethylene terephthalate), is made by the esterification of terephthalic acid with methyl alcohol, as shown in the following equation:

$$HOOC-C_6H_4-COOH + 2CH_3OH \longrightarrow CH_3OOC-C_6H_4-COOCH_3 + 2H_2O$$

The next step in the production of polyesters involves transesterification. Dimethyl terephthalate reacts with ethylene glycol to form diethyleneglycol terephthalate, as shown by the following equation:

$$CH_3OOC-C_6H_4-COOCH_3 + 2HOCH_2CH_2OH \longrightarrow HOCH_2CH_2OOC-C_6H_4-COOCH_2CH_2OH + 2CH_3OH$$

The diethyleneglycol terephthalate is polymerized to polyester for melt spinning of fibers and filaments.

11. HALOGENATION

Halogenation is the process whereby a halogen is used to introduce one or more halogen atoms into an organic compound. Reactions in which the halogenating agent is a halogen acid, such as hydrochloric acid, are listed in the table for the unit process of hydrohalogenation.

The preparation of organic compounds containing fluorine, chlorine, bromine and iodine can be accomplished by a variety of methods. The conditions and procedures differ, not only for each member of the halogen family, but also with the type and structure of the compound undergoing treatment.

The chlorine derivatives, because of the greater economy in effecting their preparation, are by far the most important of the technical halogen compounds. For this reason, they are given primary consideration in the Halogenation Table.

A process for the production of ethylene dichloride is the direct chlorination of ethylene in the presence of ferric chloride catalyst, as shown in the following equation:

$$CH_2{=}CH_2 + Cl_2 \xrightarrow[135^\circ C]{FeCl_3} ClCH_2CH_2Cl$$

Methyl chloride, or chloromethane, is made by the direct chlorination of methane, as shown by the following equation:

$$CH_4 + Cl_2 \longrightarrow CH_3Cl + HCl$$

Chlorobenzene is manufactured by passing dry chlorine into benzene in the presence of a catalyst such as $FeCl_3$, $AlCl_3$, Fuller's earth or iron turnings. The equation for the reaction follows:

$$C_6H_6 + Cl_2 \xrightarrow[40^\circ C]{\text{catalyst}} C_6H_5Cl + HCl$$

12. HYDRODEALKYLATION

Hydrodealkylation is the process by which methyl, or larger alkyl groups, are removed from hydrocarbon molecules and replaced by hydrogen atoms. A catalyst may or may not be used; when used they are usually of the chromium-on-alumina, or platinum-on-alumina, type. Hydrodealkylation is used in petroleum refining to upgrade products of low value, such as heavy reformate fractions, naphthenic crudes or recycle stocks from catalytic cracking.

Hydrodealkylation is used in the Detol process of the Houdry Division of Air Products and Chemicals, Inc., which produces high-purity benzene from toluene and/or xylenes and/or C_9 and heavier aromatics. The equation for the formation of benzene from toluene is shown below:

$$C_6H_5CH_3 + H_2 \xrightarrow[540\text{-}650^\circ C]{\text{catalyst}} C_6H_6 + CH_4$$

Naphthalene is produced from alkylnaphthalenes by the Hydeal process of UOP, Inc. In this process, a feedstock derived from heavy raffinate is processed with excess hydrogen over a chomia-alumina catalyst of high purity and low sodium content. The equation for the formation of naphthalene from methyl naphthalene follows:

$$C_{10}H_7CH_3 + H_2 \xrightarrow[\text{heat}]{\text{catalyst}} C_{10}H_8 + CH_4$$

13. HYDROGENATION

Hydrogenation is the process in which hydrogen is added to an organic compound. It may occur as a direct addition of hydrogen to the double bond of an unsaturated molecule, resulting in a saturated product. An example is the catalytic hydrogenation of benzene to cyclohexane, which may be carried out in either the vapor or liquid phase. The equation for the reaction follows:

$$C_6H_6 + 3H_2 \xrightarrow[210^\circ C]{\text{Ni or Pt catalyst}} \text{(S)}$$

In organic compounds containing nitro groups, hydrogen replaces oxygen to form amines. An example is the vapor phase heterogeneous catalytic hydrogenation of nitrobenzene to aniline using a copper catalyst. The equation for the reaction follows:

$$C_6H_5NO_2 + 3H_2 \xrightarrow[270\text{-}475^\circ C]{\text{Cu on silica}} C_6H_5NH_2 + 2H_2O$$

Hydrogen is added to aldehydes or ketones to produce alcohols. An example is the hydrogenation of *n*-butyraldehyde to *n*-butyl alcohol over a fixed-bed catalyst at 130-160°C and 30-50 atm. The equation for the reaction follows:

$$CH_3CH_2CH_2CHO + H_2 \xrightarrow[130\text{-}160^\circ C]{\text{catalyst}} CH_3CH_2CH_2CH_2OH$$

14. HYDROHALOGENATION

Hydrohalogenation is the process in which a halogen atom is added to an organic compound with a halogen acid, such as hydrogen chloride, serving as the halogenating agent. As an example, methyl chloride is produced by the action of hydrogen chloride on methanol in the vapor phase, in the presence of a catalyst. The reaction occurs at atmospheric pressure at a temperature of 340-350°C, using a catalyst such as alumina gel, zinc chloride-on-pumice, cuprous chloride or activated carbon. The equation for the reaction is:

$$CH_3OH + HCl \xrightarrow[340\text{-}350^\circ C]{\text{catalyst}} CH_3Cl + H_2O$$

15. HYDROLYSIS AND HYDRATION

Hydrolysis is a chemical process in which water reacts with another substance to form two or more new substances. Water is ionized in the hydrolysis process and the compound that is hydrolyzed is split, with hydrogen going to one of the new substances and hydroxyl to the other.

An example is the manufacture of isopropyl alcohol from propylene by sulfonation, followed by hydrolysis. In the BP Chemicals process, a solution of diisopropyl sulfate and isopropyl acid sulfate (*see* Sulfonation pg. 111) is hydrolyzed to isopropyl alcohol in a hydrolyzer-stripper in the presence of dilution water, as shown in the following equation:

$$[(CH_3)_2CH]_2SO_4 + (CH_3)_2CHOSO_2OH + 3H_2O \longrightarrow 3(CH_3)_2CHOH + 2H_2SO_4$$

Hydration is the process in which water reacts with a compound without decomposition of the compound, forming a new compound containing both the hydrogen and hydroxyl groups. An example is the Veba-Chemie AG process for the production of isopropyl alcohol by the direct hydration of propylene with demineralized water, as shown by the following equation:

$$CH_3CH{=}CH_2 + H_2O \xrightarrow[170\text{-}190^\circ C]{\text{catalyst}} (CH_3)_2CHOH$$

16. NITRATION

Nitration is the unit process in which one or more nitro groups ($-NO_2$) are introduced into organic compounds by the use of nitric acid. Aromatic nitrations are usually effected with mixed acid, a mixture of nitric acid and concentrated sulfuric acid, as shown by the following equation for the nitration of benzene:

$$C_6H_6 + HNO_3 \xrightarrow[50\text{-}90^\circ C]{H_2SO_4} C_6H_5NO_2 + H_2O$$

This is an electrophilic substitution reaction, in which the electrophilic agent is the nitronium ion, NO_2^+. The sulfuric acid protonates the nitric acid to give $H_2NO_3^+$, which loses water to form NO_2^+, as shown in the following equation:

$$H_2SO_4 + HNO_3 \rightleftharpoons HSO_4^- + H{-}\underset{\displaystyle H}{\overset{+}{O}}{-}NO_2$$

$$H{-}\underset{\displaystyle H}{\overset{+}{O}}{-}NO_2 \rightleftharpoons NO_2^+ + H_2O$$

17. OXIDATION

The unit process of oxidation of organic compounds generally means chemical reaction with oxygen to introduce one or more oxygen atoms into the compound, and/or to remove hydrogen atoms from the compound. The term "oxidation" has been broadened to include any reaction in which electrons are lost. Oxidation and reduction always occur simultaneously (redox reactions). The substance losing electrons is oxidized and the substance receiving electrons is reduced.

The oxidation process is used in the Aldehyd GmbH process for acetaldehyde production by the direct oxidation of ethylene. The process uses a catalyst solution of cupric chloride, which contains small quantities of palladium chloride. The reactions may be summarized as follows:

$$CH_2{=}CH_2 + 2CuCl_2 + H_2O \xrightarrow{PdCl_2} CH_3CHO + 2CuCl + 2HCl$$

$$2CuCl + 2HCl + \tfrac{1}{2}O_2 \longrightarrow 2CuCl_2 + H_2O$$

In the reaction, palladium chloride is reduced to elemental palladium and hydrochloric acid and is reoxidized by cupric chloride. During catalyst regeneration, the cuprous chloride is reoxidized to cupric chloride.

An example of oxidation, in which the oxygen atom is already present in the molecule and oxidation consists of removal of hydrogen atoms, is the synthesis of formaldehyde from methanol. In the Meissner process, a mixture of methanol in air is sent to a reactor filled with ferromolybdenum oxide maintained at 300°C. The equation for the reaction is as follows:

$$CH_3OH + \tfrac{1}{2}O_2 \xrightarrow[300^\circ C]{catalyst} H\overset{\overset{\displaystyle O}{\|}}{C}H + H_2O$$

Oxidation of an aromatic compound is illustrated by the production of phthalic anhydride from *o*-xylene. In the Von Heyden process for production of phthalic anhydride, air at 1.5-2 atm pressure is heated to 140-160°C, mixed with vaporized *o*-xylene, and passed over a proprietary granular catalyst. The equation for the reaction follows:

$$C_6H_4(CH_3)_2 + 3O_2 \xrightarrow[140\text{-}160^\circ C]{\text{catalyst}} C_6H_4(CO)_2O + 3H_2O$$

18. OXYHALOGENATION

In the oxyhalogenation process, halogenation is carried out by catalytically oxidizing a halogen acid to the halogen with air or oxygen. Commercial applications of the process use oxychlorination, in which chlorination is carried out by catalytically oxidizing hydrogen chloride to chlorine with air or oxygen. This process has also been termed oxyhydrochlorination.

The oxychlorination process is of particular importance in the manufacture of vinyl chloride monomer (VCM) from ethylene, chlorine, and air or oxygen. The Monsanto Company/Scientific Design Company process for the manufacture of vinyl chloride monomer utilizes a "balanced" plant, wherein all hydrogen chloride produced as a coproduct of the dehydrochlorination of ethylene dichloride (EDC) to vinyl chloride monomer is recycled to the oxychlorination reactor. The equations for the reactions are:

$$CH_2{=}CH_2 + Cl_2 \longrightarrow ClCH_2CH_2Cl \text{ (direct chlorination)}$$

$$ClCH_2CH_2Cl \longrightarrow CH_2{=}CHCl + HCl \text{ (dehydrochlorination)}$$

$$CH_2{=}CH_2 + 2HCl + \tfrac{1}{2}O_2 \longrightarrow CLCH_2CH_2Cl + H_2O \text{ (oxychlorination)}$$

19. PHOSGENATION

Phosgenation is the process in which phosgene ($COCl_2$) reacts with an amine to form an isocyanate, or an alcohol to form a carbonate.

Toluene diisocyanate is an important starting material in the manufacture of polyurethanes. The reaction for the synthesis of 2,4-toluene diisocyanate from 2,4-toluene-diamine and phosgene is:

$$CH_3C_6H_3(NH_2)_2 + 2ClCOCl \longrightarrow CH_3C_6H_3(NCO)_2 + 4HCl$$

Polycarbonate resins are made by the reaction of bisphenol A with phosgene, as shown by the following equation:

$$HO\text{-}C_6H_4\text{-}C(CH_3)_2\text{-}C_6H_4\text{-}OH + nClCOCl \xrightarrow{\text{pyridine}} H\left[O\text{-}C_6H_4\text{-}C(CH_3)_2\text{-}C_6H_4\text{-}OC(O)\right]_n Cl + (2n\text{-}1)HCl \quad \text{pyridine}$$

20. POLYMERIZATION

Polymerization is a chemical process in which a large number of relatively simple molecules combine to form a chain-like macromolecule. The reaction is usually carried out with a catalyst, often under high pressure.

Industrial polymerizations are performed by subjecting unsaturated or other reactive compounds to conditions that will bring about combination. Polymerizations generally can be divided into two types: (1) condensation polymerizations, in which polymers are formed by the elimination of small molecules, such as water; and (2) addition polymerizations, in which saturation of one or more unsaturated bonds is effected. For the manufacture of nylon 66, an equimolar mixture of adipic acid and hexamethylenediamine is heated at 270°C and 10 atm pressure. The equation for the reaction follows:

$$nHO\overset{O}{\overset{\|}{C}}(CH_2)_4\overset{O}{\overset{\|}{C}}OH + nNH_2(CH_2)_6NH_2 \rightarrow \left[NH\overset{O}{\overset{\|}{C}}(CH_2)_4\overset{O}{\overset{\|}{C}}NH(CH_2)_6\right]_n + 2nH_2O$$

Addition polymerization occurs in the manufacture of polyvinyl chloride from vinyl chloride. The polymerization can be carried out in a water suspension, which contains a soap as an emulsifier and a persulfate initiator. The equation for the reaction follows:

$$nCH_2{=}\underset{Cl}{CH} \longrightarrow (CH_2{-}\underset{Cl}{CH})_n$$

21. PYROLYSIS

Pyrolysis is the unit process in which the chemical change of a substance occurs by heat alone. Pyrolysis includes thermal rearranagements into isomers, thermal polymerizations and thermal decompositions. Pyrolysis does not include thermal changes that require catalysts, or changes that are initiated by other forms of energy, such as ultraviolet radiation.

A major use of the pyrolysis process is in the production of polymer grade ethylene (99.9% pure) by the steam pyrolysis of hydrocarbons. Major by-products include propylene, butadiene and C_6-C_8 aromatics-rich pyrolysis gasoline. In the C-E Lummus process, the hydrocarbon feedstock is preheated and cracked in the presence of steam in tubular short residence time (SRT) pyrolysis furnaces at 760-870°C. The equation for the formation of ethylene from ethane is:

$$CH_3CH_3 \xrightarrow[760\text{-}870^\circ C]{} CH_2 \quad CH_2 + H_2$$

22. REFORMING (STEAM)-WATER-GAS REACTION

Steam reforming is the process in which steam reacts with methane or other hydrocarbons to form hydrogen and either carbon monoxide or carbon dioxide, as shown below:

$$CH_4 + H_2O \xrightarrow[800^\circ C]{catalyst} CO + 3H_2$$

$$CH_4 + 2H_2O \xrightarrow[800^\circ C]{catalyst} CO_2 + 4H_2$$

Since the reaction mixture is deficient in carbon for subsequent methanol synthesis, carbon dioxide is admixed with the methane and steam prior to forming. This promotes the reverse water-gas shift reaction, as shown below:

$$H_2 + CO_2 \xrightarrow[800^\circ C]{catalyst} CO + H_2O$$

The basic process steps in the manufacture of synthesis gas from natural gas (methane), steam and carbon dioxide comprise preheating the reactants; passing them through externally heated, catalyst-packed tubes (the catalyst generally consists of promoted nickel); conducting the reaction mixture through a waste heat boiler, and then through the heat exchangers in which the incoming reactants are preheated.

Methanol is produced in a medium pressure process by the catalytic reduction of synthesis gas which consists of two parts of hydrogen and one part of carbon monoxide at 100-350 atm pressure and 250-400°C. The reaction is shown by the following equation:

$$2H_2 + CO \xrightarrow[250\text{-}400^\circ C]{catalyst} CH_3OH$$

The Imperial Chemical Industries (ICI) low-pressure process for the synthesis of methanol uses gaseous liquid or solid hydrocarbon feedstocks. The hydrocarbon feedstock is desulfurized by catalytic or adsorptive processes, mixed with process steam and passed to a tubular reformer. The reformed gas is converted to methanol in a pressure vessel containing a single bed of a stable, copper-based methanol synthesis catalyst operating in the pressure range of 50-100 atm at 200-300°C.

23. SULFONATION

Sulfonation is the process by which the sulfonic acid group ($-SO_2OH$), or the corresponding salt, or sulfonyl halide, is attached to a carbon atom. Sulfonation is also used to indicate treatment of any organic compound with sulfuric acid, regardless of the nature of the products formed. The Sulfonation Table includes products made by sulfation, which involves the placement of an $-OSO_2OH$ group on carbon when sulfating an alkene, or of an $-SO_2OH$ group on oxygen when sulfating an alcohol or phenol.

The indirect hydration method for the production of isopropyl alcohol uses the sulfation of propylene to isopropyl hydrogen sulfate and subsequent hydrolysis to isopropyl alcohol and sulfuric acid. The equations for the reactions follow:

$$CH_3CH{=}CH_2 + H_2SO_4 \longrightarrow (CH_3)_2CHOSO_2OH$$

$$(CH_3)_2CHOSO_2OH + H_2O \longrightarrow (CH_3)_2CHOH + H_2SO_4$$

The sulfonation of "detergent alkylate" or mixed linear alkylbenzenes, in which the alkyl group averages about C_{12}, is important for the large-scale manufacture of household and industrial detergents.

Detergent alkylate is sulfonated commercially with SO_3 or oleums of various strengths. Vaporized SO_3 gives the least dealkylation of the product and the best odor. The Chemithon Corporation process uses SO_3 vapor diluted with a carrier gas in a continuous operation, with reaction and heat removal occurring in a thin film on a cooled reactor surface. Detergent alkylate forms almost entirely the *p*-sulfonic acid, as shown by the following equation:

$$R{-}C_6H_5 + SO_3 \longrightarrow p\text{-}R{-}C_6H_4{-}SO_2OH$$

5

DEFINITIONS OF MINOR UNIT PROCESSES

1. ACID CLEAVAGE

Acid cleavage is the process in which an organic chemical is cleaved into two or more compounds by the action of an acid catalyst. Acid cleavage is used in the production of phenol and acetone from cumene by air oxidation to cumene hydroperoxide, which is concentrated and cleaved to phenol and acetone by acid catalysis.

In the BP Chemicals International Ltd. and Hercules Inc. process for the production of phenol and acetone, cumene is oxidized with air under carefully controlled conditions to produce cumene hydroperoxide at high efficiency. The oxidate is concentrated to about 80% cumene hydroperoxide in special equipment prior to cleavage with an acid catalyst. The cleavage reactor is of special design to combine high efficiency and safety. After removal of the acid catalyst, the cleavage mixture is fractionated to isolate and purify the acetone and phenol products. The equations for the reactions follow:

$$C_6H_5{-}CH(CH_3)_2 \xrightarrow[130^\circ C]{air} C_6H_5{-}C(CH_3)_2{-}OOH$$

$$C_6H_5{-}C(CH_3)_2{-}OOH \xrightarrow[55\text{-}65^\circ C]{10\text{-}25\%\ H_2SO_4} C_6H_5OH + CH_3\overset{O}{\overset{\|}{C}}CH_3$$

2. ACID REARRANGEMENT

An acid rearrangement is a chemical reaction in which the atoms of a single compound recombine under the influence of an acid catalyst to form a new compound having the same molecular weight but different properties.

The production of *p*-aminophenol is carried out by suspending nitrobenzene in sulfuric acid and treating with hydrogen in the presence of molybdenum disulfide. The nitrobenzene is reduced to, N-phenylhydroxylamine at 150°C, with hydrogen pressure of 23-33 atm. The N-phenylhydroxylamine undergoes immediate rearrangement under the influence of the sulfuric acid catalyst to form *p*-aminophenol. The equations for the reaction follow:

$$C_6H_5\text{-}NO_2 + 2H_2 \xrightarrow[150^\circ C]{MoS_2} [C_6H_5\text{-}NHOH] + H_2O$$

$$[C_6H_5\text{-}NHOH] \xrightarrow[150^\circ C]{H_2SO_4} HO\text{-}C_6H_4\text{-}NH_2$$

3. AMINATION BY REDUCTION

Amination by reduction is the process of forming an amine from a compound that already contains a carbon-to-nitrogen bond. Reduction may occur in the following ways: (1) acceptance of one or more electrons by an atom or ion; (2) removal of oxygen from a compound; (3) addition of hydrogen to a compound. In this Guide, those processes using hydrogen in the presence of a catalyst to prepare amines are included in the Hydrogenation Table.

Reduction of nitro compounds with iron and hydrochloric acid, termed the Bechamp reduction, has been of considerable commercial importance. Presently, the manufacture of aniline from nitrobenzene and other large-scale reductions is carried out by catalytic hydrogenation. However, the iron reduction process is still used for smaller-scale reductions, such as the production of *o*-aminophenol from *o*-nitrophenol. The equations for the reactions follow:

$$NO_2\text{-}C_6H_4\text{-}OH + 3Fe + 4H_2O \xrightarrow[HCl]{FeCl_2} NH_2\text{-}C_6H_4\text{-}OH + Fe(OH)_2 + FeO + Fe(OH)_3 + (H)$$

$$Fe(OH)_2 + 2Fe(OH)_3 \longrightarrow Fe_3O_4 + 4H_2O$$

The reduction of 1-chloro-2-nitrobenzene with zinc and strong alkali yields 2,2′-dichlorohydrazobenzene, as shown in the section entitled Benzidene Rearrangement.

4. BECKMANN REARRANGEMENT

The Beckmann rearrangement is the conversion of oximes of ketones (ketoximes) into substituted acid amides by means of acidic reagents. The generic equation follows:

$$R{-}\underset{\underset{HON}{\|}}{C}{-}R' \xrightarrow{PCl_5} R{-}\overset{\overset{O}{\|}}{C}{-}NHR'$$

An important commercial use of this process is the rearrangement of cyclohexanone oxime to caprolactam, an intermediate in the manufacture of a synthetic fiber, nylon 6. The equation for the reaction follows:

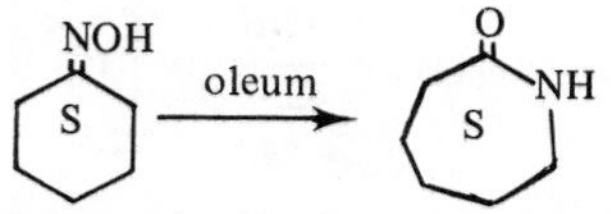

$$\text{cyclohexanone oxime (=NOH, ring S)} \xrightarrow{\text{oleum}} \text{caprolactam (C=O, NH, ring S)}$$

The Stamicarbon BV process for production of caprolactam from cyclohexanone, which uses the Beckmann rearrangement, is described in the section entitled Oximation, pg. 121.

5. BENZIDINE REARRANGEMENT

This refers to the rearrangement of hydrazobenzenes in the presence of strong acids to form benzidines (4,4′-diaminobiphenyls). Benzidine dyes are a group of azo dyes derived from 3,3′-dichlorobenzidine, which are made by the benzidine rearrangement of the intermediate 2,2′-dichlorohydrazobenzene, as shown by the following equation:

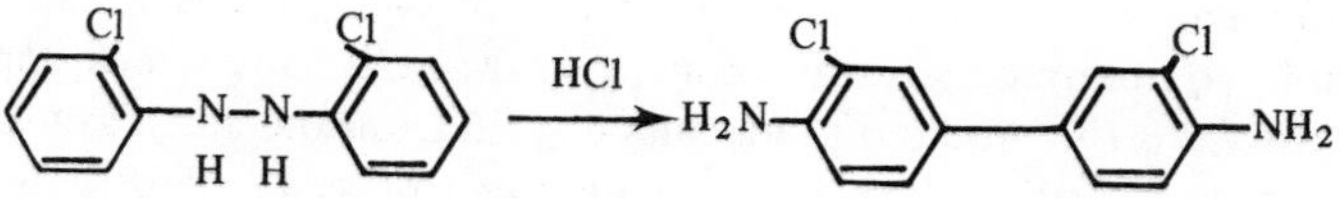

$$ClC_6H_4{-}\underset{H}{N}{-}\underset{H}{N}{-}C_6H_4Cl \xrightarrow{HCl} H_2N{-}C_6H_3Cl{-}C_6H_3Cl{-}NH_2$$

The intermediate, 2,2′-dichlorohydrazobenzene, can be made by the reduction of 1-chloro-2-nitrobenzene with zinc and alkali, as shown by the following equation:

$$2\,ClC_6H_4NO_2 + 5Zn + 10NaOH \longrightarrow ClC_6H_4{-}\underset{H}{N}{-}\underset{H}{N}{-}C_6H_4Cl + 5Na_2ZnO_2 + 4H_2O$$

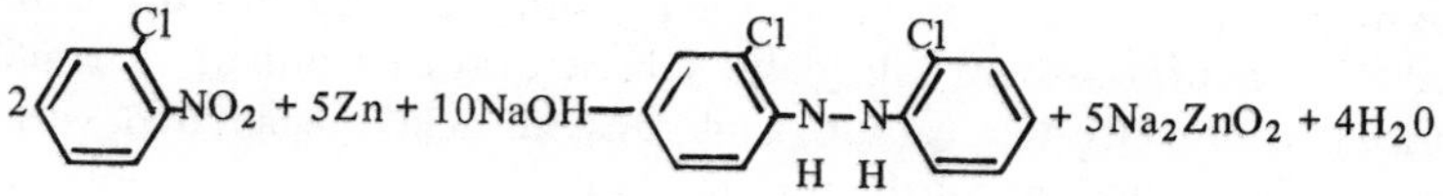

6. CANNIZZARO REACTION

In the Cannizzaro reaction, aldehydes without α-hydrogen atoms undergo self-oxidation-reduction in the presence of strong alkali to yield an organic acid and an alcohol. When the aldehydes are not identical, the reaction is called a "crossed Cannizzaro reaction."

The crossed Cannizzaro reaction is a part of the Josef Meissner GmbH process for the continuous preparation of pentaerythritol from acetaldehyde and formaldehyde. The initial step is an aldol condensation between three moles of formaldehyde and acetaldehyde, in an aqueous solution of calcium hydroxide at 15-45°C. The intermediate, tris(hydroxymethyl)acetaldehyde, is formed as shown by the following equation:

$$3HCHO + CH_3CHO \xrightarrow[15\text{-}45^\circ C]{Ca(OH)_2} \left[HOCH_2-\underset{CH_2OH}{\overset{CH_2OH}{\overset{|}{\underset{|}{C}}}}-CHO \right]$$

The next step is a Cannizzaro reaction between the intermediate tris (hydroxymethyl)acetaldehyde and another mole of formal dehyde, which results in reduction of the intermediate to pentaerythritol and the oxidation of the formaldehyde to formic acid, as shown below:

$$\left[HOCH_2-\underset{CH_2OH}{\overset{CH_2OH}{\overset{|}{\underset{|}{C}}}}-CHO \right] + HCHO \xrightarrow{Ca(OH)_2} HOCH_2-\underset{CH_2OH}{\overset{CH_2OH}{\overset{|}{\underset{|}{C}}}}-CH_2OH + HCOOH$$

7. CARBOXYLATION

Carboxylation is the process by which a carboxyl group is introduced into an organic compound to form a carboxylic acid. A commercial application of the process is the production of salicylic acid from the alkali salt of phenol by the action of carbon dioxide under heat and pressure. This is also known as the Kolbe-Schmidt reaction.

In the commercial process to produce salicylic acid, the sodium phenolate solution is introduced into a revolving heated ball mill. The mill is evacuated and heated to 130°C to reduce the sodium phenolate to a fine, dry powder. Carbon dioxide is added under 7 atm pressure, and the mixture is heated at 100°C to form sodium salicylate. This is dissolved out of the mill, decolorized with activated carbon, and salicylic acid precipitated with sulfuric acid. The equations for the reactions are:

$$2\, C_6H_5ONa + CO_2 \xrightarrow[100^\circ C]{7\ atm} C_6H_4(ONa)COONa + C_6H_5OH$$

$$C_6H_4(ONa)COONa + H_2SO_4 \longrightarrow C_6H_4(OH)COOH + Na_2SO_4$$

8. CHLOROHYDRINATION

This process involves the addition of hypochlorous acid to the double bond of an olefin to give a chlorohydrin in, which a hydroxyl group and a chlorine atom are attached to adjacent carbon atoms. Chlorohydrination is used in the Shell process to produce glycerin from propylene. In this process, allyl chloride is treated with hypochlorous acid at 38°C to produce glycerin dichlorohydrin, as shown by the following equation:

$$CH_2{=}CHCH_2Cl + HOCl \xrightarrow{38^\circ C} CH_2(Cl){-}CH(OH){-}CH_2(Cl)$$

The glycerin dichlorohydrin can be hydrolyzed directly to glycerin with two moles of caustic soda, or to the intermediate epichlorohydrin, using the cheaper calcium hydroxide, as shown in the following equations:

$$CH_2(Cl){-}CH(OH){-}CH_2(Cl) \xrightarrow[H_2O]{Ca(OH)_2} \underset{\text{(epichlorohydrin: } H_2C{-}O{-}CH{-}CH_2Cl)}{} \xrightarrow[H_2O]{NaOH} CH_2OH{-}CHOH{-}CH_2OH$$

9. ELECTROHYDRODIMERIZATION

Electrohydrodimerization is the process in which an organic compound undergoes reductive dimerization by direct electrolysis. The only example in the table is the production of adiponitrile by the direct electrohydrodimerization of acrylonitrile. A description of the Asahi process follows:

Acrylonitrile is fed to the catholyte tank, where it is emulsified with supporting salt. The resulting emulsion is sent to the electrolyzer and constantly recirculated through the electrolyzer and the catholyte tank. A portion of acrylonitrile dissolved in the catholyte is converted into adiponitrile by an electrohydrodimerization reaction on the cathode surface. This process utilizes cation exchange membranes, which migrate cations selectively to prevent oxidation of the reaction mixture in the anode compartment, as shown in the following diagram:

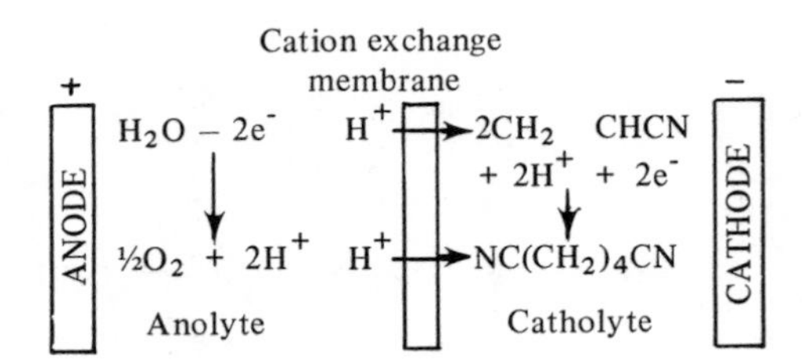

H^+ ions to be consumed in the dimerization reaction migrate through the cation exchange membrane. The supporting salt is of a simple formula, such as tetraethyl ammonium sulfate, and has a low affinity to organic materials. Therefore, the supporting salt can easily be separated from the oily layer.

A portion of the catholyte is sent to the acrylonitrile stripper, where unconverted acrylonitrile and by-products such as propionitrile are removed, together with a small quantity of water from the top. The top stream is separated into two layers in the decanter. Oily effluent is sent to the propionitrile stripper, where acrylonitrile is recovered from the bottom. Aqueous effluent is sent to the water stripper, where dissolved acrylonitrile is recovered together with the by-products from the top. Water is removed from the bottom.

The bottom stream of the acrylonitrile stripper is separated into two layers in the coalescer. The aqueous effluent is recycled to the catholyte tank through the catholyte purification unit. After removing dissolved water in the flush chamber, the oily effluent is sent to the heavy-cut column, where high-boiling components, such as the trimer of acrylonitrile, are removed from the bottom. The top stream of the heavy-cut column is sent to the first light-cut column, where low-boiling components are removed from the top and adiponitrile is obtained from the lower part of the column.

Adiponitrile is recovered from the bottom stream of the heavy-cut column and the top stream of the first light-cut column for recycling to the process.

10. EPOXIDATION

Epoxidation is the process by which an epoxide is formed. In epoxidation, an alkene reacts with a peracid, the pi bond of the alkene is broken, and a three-membered cyclic ether called an epoxide, or oxirane, is formed. The generic equation for the reaction is:

$$\gt C{=}C\lt \; + \; R\overset{O}{\overset{\|}{C}}{-}O{-}OH \longrightarrow \gt C\overset{O}{\frown}C\lt \; + \; R\overset{O}{\overset{\|}{C}}{-}OH$$

A peracid — An expoxide (oxirane)

The Daicel process produces propylene oxide and acetic acid by expoxidizing propylene with peracetic acid as follows: Propylene and a solution of 30% peracetic acid and 10-15% acetic acid in ethyl acetate containing stabilizer is continuously fed into a series of 3 specially designed reactors for epoxidation at 50-80°C under 9-12 atm pressure. In about 2-3 hours, 97-98% of the peracetic acid is reacted to give propylene oxide in 90-92% yield. Then the reaction products are fed into a stripper and distilled at 1.3-5.2 atm pressure. From the top of the stripper, a mixture of propylene and propylene oxide and, from its bottom, a mixture of ethyl acetate and acetic acid, are discharged.

The first mixture is liquefied by cooling and compression, then fed into a propylene stripping column, operated at 12-15 atm pressure, where propylene is recovered and recylced into the first reactor. Crude propylene oxide is discharged from the bottom of the propylene column into a light-end column and a propylene oxide column for refining.

The second mixture and a high-boiler, which is obtained from the bottom of the propylene oxide column, are fed into recovery columns, where ethyl acetate is recovered and recycled into the peracetic acid reactor. Acetic acid is also recovered and used as is, or refined. The equation for the reaction follows:

$$CH_3CH{=}CH_2 + CH_3\overset{\overset{O}{\|}}{C}OOH \rightarrow CH_3{-}\underset{\diagdown O \diagup}{CH{-}CH_2} + CH_3\overset{O}{\overset{/\!/}{C}}OH$$

11. HYDROACETYLATION

The hydroacetylation process is the addition of acetic acid to acetylene. When acetylene and acetic acid are reacted using a suitable catalyst, union takes place to form vinyl acetate or the acetate ester of ethylidine glycol, as shown below:

$$HC{\equiv}CH + CH_3COOH \rightarrow CH_3COOCH{=}CH_2$$

$$HC{\equiv}CH + CH_3COOH \rightarrow CH_3CH(O\overset{O}{\overset{/\!/}{C}}CH_3)_2$$

The usual catalysts are strong acids, such as sulfuric acid, phosphoric acid (with or without mercury salts), boron fluoride and salts of various metals.

The process may be illustrated by the following description of a vapor-phase reaction between acetylene and acetic acid in the presence of a zinc acetate catalyst to yield vinyl acetate:

Acetylene is purified to remove hydrogen sulfide and phosphorus compounds. It is then mixed in slight excess with vaporized acetic acid and fed to a multitubular, fixed-bed reactor, which contains a catalyst of zinc acetate deposited on activated carbon (10% zinc). The reaction is exothermic, so the reactor is cooled by circulating oil around the tubes. The reactor temperature is maintained at 175-200°C. The reactor effluent is condensed and fed to a light-ends column, where acetylene, methyl acetylene, propadiene and other light-ends are removed from the top of the column. The acetylene must be repurified before it may be recycled.

Vinyl acetate product is distilled overhead in a vinyl acetate column. Recycle acetic acid is separated from heavy-ends in a recovery column.

12. HYDROCYANATION

The hydrocyanation process is the addition of hydrogen cyanide to an organic compound. Hydrogen cyanide reacts with carbonyl compounds to form a cyanohydrin, as shown by the following generic equation:

$$R\overset{O}{\overset{\|}{C}}R' + HCN \rightarrow R - \overset{HO\quad CN}{\overset{\backslash\ /}{C}} - R'$$

Hydrogen cyanide will also react with alkynes, as shown by the following reaction with acetylene to form acrylonitrile:

$$HC \equiv CH + HCN \rightarrow CH_2 = CHCN$$

Hydrocyanation is the first step in the production of methyl methacrylate. Acetone is reacted with hydrogen cyanide in the presence of a base to form acetone cyanohydrin, as shown by the following equation:

$$CH_3\overset{O}{\overset{\|}{C}}CH_3 + HCN \xrightarrow[\text{base}]{} CH_3\overset{HO\quad CN}{\overset{\backslash\ /}{C}}CH_3$$

Acetone cyanohydrin is reacted with sulfuric acid to form methacrylamide sulfate, which is further hydrolyzed and esterified in a continuous process to form methyl methacrylate.

13. ISOMERIZATION

Isomerization is the process whereby an organic compound under the influence of heat and a catalyst is converted to another compound having different properties, by changing the arrangement of atoms in the molecule without changing the number of atoms. Isomerization is used in petroleum refining to convert straight-chain hydrocarbons into branched-chain hydrocarbons of the same molecular weight, as in conversion of n-butane to isobutane, which is needed for alkylation feedstock. Isomerization catalysts include aluminum chloride, antimony chloride, platinum, other metals and metal compounds. Reaction conditions vary according to the catalyst employed and the feedstock processed, ranging from 400-480°C and 7-50 atm pressure.

An example of isomerization is the Isomar process of UOP, Inc. for production of *p*-xylene and/or *o*-xylene from C_8 aromatics. In this process, any nonequilibrium mixture of C_8 aromatics is efficiently isomerized toward equilibrium. The description of the Isomar process follows:

The feedstock may be any C_8 aromatic mixture, *e.g.,* from catalytic reformates or from pyrolysis gasoline. The latter often contains as much as 40% ethylbenzene, which is no detriment to the operation. Feedstocks may also be pure solvent extracts or fractional heart-cuts, containing as much as 25% saturates. Hydrogen supply may be from catalytic reforming or any other suitable source. Chemical hydrogen consumption is minor.

C_8 aromatic reactor feed, deficient in one or more components relative to equilibrium composition, is processed over a fixed bed of catalyst in the presence of hydrogen. The liquid portion of the effluent is fractionated to remove both light and heavy aromatic by-products, as well as the cracked materials resulting from inclusion of saturates in the feed. *p*-xylene and/or *o*-xylene are separated from the resulting fractionation heart-cut, thereby preparing a recycle material to the Isomar reactor. Fresh feedstock is introduced into the circuit in the most appropriate place. The general equation for the reaction is:

$$C_8 \text{ aromatic mixture} \xrightarrow[\text{heat}]{\text{catalyst}} C_6H_4(CH_3)_2 \ (o\text{-xylene}) + CH_3C_6H_4CH_3 \ (p\text{-xylene})$$

14. OXIMATION

This is the process of introducing an oxime group, C=NOH, into an organic compound. Oximes are formed by the reaction of hydroxylamine, or its acid salts, with the carbonyl group of aldehydes, ketones or quinones. The generic equation follows:

$$-\overset{\overset{O}{\|}}{C}- + NH_2OH \longrightarrow \ \rangle C{=}NOH + H_2O$$

Oximation is a step in the Stamicarbon BV process for the production of high-purity fiber-grade caprolactam, starting from cyclohexanone, ammonia and hydrogen. The hydroxylamine is produced in a gas-liquid contactor from nitrate ions (obtained by the oxidation of ammonia) and hydrogen in the presence of a noble metal catalyst. The nitrate ions are reduced to hydroxylamine in the presence of a buffering acid, as shown by the following equation:

$$NO_3^- + 2H^+ + 3H_2 \longrightarrow NH_3OH^+ + 2H_2O$$

The formed hydoxylamine reacts with pure cyclohexanone in the presence of an organic solvent at a pH below 4. The equation for the reaction follows:

$$\text{(S)=O} + NH_2OH \longrightarrow \text{(S)=NOH} + H_2O$$

The cyclohexanone oxime is separated from the solvent by distillation and pumped to the Beckmann rearrangement unit, where heating with oleum causes the rearrangement to caprolactam (*see* Beckmann Rearrangement, pg. 115). The rearrangement mixture is neutralized with aqueous ammonia and, after settling, the crude caprolactam solution is decanted from the ammonium sulfate solution for further purification.

15. OXYACETYLATION

Oxyacetylation is the process in which oxygen and the acetyl group $CH_3\overset{O}{\overset{\|}{C}}-$ are added to an olefin to produce an unsaturated acetate ester. Oxyacetylation is used in the new commercial process to make vinyl acetate from ethylene, acetic acid and oxygen. The equation for the reaction follows:

$$CH_3\overset{O}{\overset{\|}{C}}OH + CH_2{=}CH_2 + \tfrac{1}{2}O_2 \longrightarrow CH_3\overset{O}{\overset{\|}{C}}OCH{=}CH_2 + H_2O$$

The U.S.I. Chemicals Division of National Distillers and Chemical Corporation process for making vinyl acetate monomer from ethylene, acetic acid and oxygen follows:

A fixed-bed tubular reactor contains a noble metal catalyst, supported on a special carrier. The feed mixture of ethylene, acetic acid and oxygen circulates through the catalyst-filled tubes. Effluent from the absorber is combined with condensed reactor effluent and fed to the primary separator facilities. At the same time, absorber off-gas is water-scrubbed to reclaim acetic acid and prevent corrosion of the recycle gas compressor.

Scrubber bottoms are combined with absorber bottoms to feed the purification process, while unreacted ethylene and oxygen saturated with water in the overhead section are sent to the carbon dioxide removal section with a compressor that provides the pressure drive in the total synthesis loop.

Carbon dioxide is the only component, aside from minor impurities that are purged, that is not totally recycled. Carbon dioxide is removed with a potassium carbonate system compatible with the vinyl acetate process. The absorber overhead, which is primarily ethylene, is fed to the acid tower for eventual return to the reactor.

The primary distillation system is made up of a primary tower and two auxiliary strippers combined with a common condenser and receiver. In the primary tower, the vinyl acetate, dissolved gases, light and heavy impurities (lower boiling than acetic acid), and water are taken overhead. The bottom stream, containing acetic acid and minor amounts of heavy impurities, is returned to the absorber and to the acid vaporizer.

The overhead condenses and separates into two layers–organic and aqueous–in a decanter. Part of the vinyl acetate layer is returned to the column as reflux to entrain more water from the column. The water layer is removed from the distillation system, and net vinyl acetate production is recovered from the system via the overhead system. The water phase is removed from the decanter and fed to the top of a tower, which strips vinyl acetate out of the aqueous bottom stream to a level of less than 10 ppm.

Vinyl acetate from the bottom of the drying tower is pumped to the lights tower, in which residual acetaldehyde and other light ends are concentrated and removed in the overhead. In the product tower, heavy impurities in the vinyl acetate are concentrated as a bottoms fraction. The tower overhead, consisting of purified vinyl acetate, is condensed, cooled and pumped to storage. Less than 25 ppm of ethyl acetate is contained in the product vinyl acetate.

16. OZONOLYSIS

Ozonolysis is the oxidation of an organic compound by means of ozone. Ozonolysis occurs readily between alkenes and ozone to cleave the double bond. The product obtained is called an ozonide, as shown in the following generic reaction:

$$>C=C< \xrightarrow{O_3} >C(-O-)(-O-O-)C<$$

Ozonides are explosive and are seldom isolated. They are decomposed by reducing agents to form aldehydes and/or ketones, or by oxidizing agents to form acids.

The Emery Industries process for the ozonolysis of oleic acid is carried out in two steps. First, an approximately 1:1 mixture of oleic acid and pelargonic acid is introduced continuously into a reactor through which is passed a countercurrent stream of oxygen containing about 2% ozone. The slightly exothermic reaction is carried out at 25-40°C. Pelargonic acid is added as a diluent to reduce the viscosity of the ozonolysis product solution. The solution is then introduced continuously into another reactor, where the ozonolysis products are oxidized at about 95°C to azelaic and pelargonic acids by a countercurrent stream of oxygen containing traces of ozone. The general reactions are:

$$CH_3(CH_2)_7{=}CH(CH_2)_7\,COOH \xrightarrow{O_3} \left[CH_3(CH_2)_7-\underset{\mid}{CH}\overset{O}{\frown}\underset{\mid}{HC}\,(CH_2)_7COOH \;\; (O-O) \right]$$

$$\xrightarrow[95^\circ C]{O_2}$$

$$\underset{\text{azelaic acid}}{HOOC\,(CH_2)_7COOH} \quad + \quad \underset{\text{pelargonic acid}}{CH_3(CH_2)_7COOH}$$

6

DIRECTORY OF COMPANIES OWNING AND/OR LICENSING DESCRIBED PROCESSES

Aicello Chemical Co., Ltd.
183 Minami-cho, Maeda
Toyohashi-shi Aichi-ken, 440
JAPAN

Airco, Inc.
575 Mountain Avenue
Murray Hill, NJ 07974

Akita Petrochemicals Co.
c/o Sumitomo Chemical Co.
Tokyo Office, 1-3-2 Marunouchi
Chiyoda-ku, Tokyo 100
JAPAN

Aldehyd GmbH
Licensor:
Hoechst-Udhe Corp.
560 Sylvan Avenue
Englewood Cliifs, NJ 07632
and
Udhe GmbH
Deggingstrasse 10-12
4600 Dortmund 1
FEDERAL REPUBLIC OF GERMANY

Allied Chemical
Columbia Road & Park Avenue
Morristown, NJ 07960

Alusuisse
Licensor:
Krupp Chemianlagenbau
Limbecker Platz 1
Essen
FEDERAL REPUBLIC OF GERMANY

American Synthetic Rubber Co.
P.O. Box 260
Louisville, KY 40201

ANIC
Licensor:
SNAM Progetti S.p.A.
20097 San Donato
Milan, ITALY

Asahi Chemical Industries Co., Ltd.
Licensor:
Asahi/America
425 Riverside Avenue
Medford, MA 02155

Atlantic-Englehard
Atlantic Richfield Co.
ARCO Technology Inc., Division
1500 Market Street
Philadelphia, PA 19101

Atlantic Richfield Co.
ARCO Technology Inc. Division
1500 Market Street
Philadelphia, PA 19101

ATO
Tour Aquitaine
Cedex No. 4
92080 Paris-La Defense
FRANCE

The Badger Company, Inc.
One Broadway
Cambridge, MA 02142

Badische-Anilin & Soda Fabrik AG
Girokonto 545 07300
6700 Ludwigshaven
FEDERAL REPUBLIC OF GERMANY

Bakol-Scientific Design
Scientific Design Co., Inc.
Two Park Avenue
New York, NY 10016

Bayer AG
509 Leverkusen, Bayerwerk
FEDERAL REPUBLIC OF GERMANY

Beaunit Corp.
P.O. Box 12234
Research Triangle Park, NC 27709

Biazzi
Licensor:
Chemical Construction Co. (CHEMICO)
1 Penn Plaza
New York, NY 10001

Blaw-Knox Company
One Oliver Plaza
Pittsburgh, PA 15222

Borden Chemical Co.
680 Fifth Avenue
New York, NY 10019

Bowmans Chemical
Licensor:
Scientific Design Co., Inc.
Two Park Avenue
New York, NY 10016

BP Chemicals International Ltd.
Devonshire House
Mayfair Place
Piccadilly, London W1X 6AY
ENGLAND

C. F. Braun & Co.
Alhambra, CA 91802

British Gas Corp.
326 High Holborn
London WC1V 7PT
ENGLAND

British Gas Corp.
North West Region
Manchester, M32 ONJ
UNITED KINGDOM

Chemical Construction Corp.
(CHEMICO)
1 Penn Plaza
New York, NY 10001

The Chemithon Corp.
5430 W. Marginal Way, S.W.
Seattle, WA 98106

Chevron Research Company
200 Bush Street
San Francisco, CA 94104

C'd F Chimie-IFP-Societe
Chimique des Charbonnages
Licensor:
Institut Francais du Petrole
1 et 4, Avenue de Bois-Preau
92-Rueil-Malmaison
FRANCE

Chiyoda Chemical Engineering &
Construction Co.
1580 Tsurumi-cho Tsurumi-ku
Yokohama-shi, Kanagawa-ken 230
JAPAN

Ciba-Geigy Corp.
Saw Mill River Road
Ardsley, NY 10502

Conoco Chemical Company
Plaza E
Saddlebrook, NJ 07662

Cosden Oil & Chemical Co.
P.O. Box 1311
Big Spring, TX 79720

Crawford & Russell Inc.
Stamford, CT 06904

Daicel, Ltd.
Toranomon, Mitsui Bldg.
8-1 Kasumigaseki 3-Chome
Chiyoda-ku
Tokyo 100
JAPAN

Daikin Kogyo Co., Ltd.
Shin-Hankyo Bldg.
8, Umeda, Kita-ku, Osaka
JAPAN

Dart Industries, Inc.
8480 Beverly Boulevard
Los Angeles, CA 90048

Denki Kagaku
Scientific Design Co., Inc.
Two Park Avenue
New York, NY 10010

Deutsche Texaco AG
Licensor:
Texaco Development Corp.
135 East 42nd Street
New York, NY 10017

Diamond Shamrock Chemical Co.
1100 Superior Avenue
Cleveland, OH 44114

Dynamit Nobel AG
Troisdorf
FEDERAL REPUBLIC OF
GERMANY

Eastman Kodak Co.
Tennessee Eastman
Kingsport, TN 37662

Englehard Industries Div.
Englehard Minerals & Chemicals Corp.
430 Mountain Avenue
Murray Hill, NJ 07974

Ethylene Plastique
Tour Aurore–Cedex No. 5
92080 Paris-La Defense
FRANCE

Fluor Engineers & Constructors, Inc.
2500 South Atlantic Boulevard
Los Angeles, CA 90040

FMC Corporation
Chemical Research & Development Center
Box 8
Princeton, NJ 08540

Foster Wheeler Corp.
110 South Orange Avenue
Livingston, NJ 07039

Gelsenberg Chemie GmbH
Rosastrasse 2
43 Essen 1 Postfach 30
FEDERAL REPUBLIC OF GERMANY

C & I/Girdler, Inc.
P.O. Box 174
1721 South Seventh Street
Louisville, KY 40201

B. F. Goodrich Chemical Co.
61 Oak Tree Blvd.
Cleveland, OH 44131

Gulf Oil Chemicals Co.
New Business Development Div.
P.O. Box 2100
Houston, TX 77001

Gulf Research & Development Co.
P.O. Drawer 2038
Pittsburgh, PA 15230

Haldor Topsoe
Licensor:
Arthur G. McKee & Co.
Cleveland, OH 44131

Hercules Inc.
910 Market Street
Wilmington, DE 19899

HIAG
Licensor:
American Lurgi Corp.
5 East 42nd Street
New York, NY 10017

Hodogaya Chemical Co., Ltd
No. 2-1 Shiba Kotoshira-cho, Minato-ku
Tokyo
JAPAN

Hoechst AG
Licensor:
Hoechst Uhde Corp.
560 Sylvan Avenue
Englewood Cliffs, NJ 07632

Honshu Chemical Industry Co.
Maruzen Bldg., 2-3-10
Nihonbashi, Chuo-ku
Tokyo 103
JAPAN

Hooker Chemical Corp.
P.O. Box 189
Niagara Falls, NY 14302

Houdry Division
Air Products & Chemical, Inc.
Box 538
Allentown, PA 18105

Hydrocarbon Research, Inc.
115 Broadway
New York, NY 10006

Idemitsu Kosan Co.
3-1-1 Marunouchi, Chiyoda-ku
Tokyo 100
JAPAN

Imperial Chemical Industries, Ltd.
Agricultural Div.
P.O. Box No. 1
Billingham, Teesside
UNITED KINGDOM

Institut Francais du Petrole
1 et 4 Avenue de Bois-Preau
92 Rueil Malmaison
FRANCE

Inventa AG
Stampfenbachstrasse 38
Zurich 6
SWITZERLAND

Japan Catalytic Chemical Industry Co.
Licensor:
Simon-Carves Ltd.
Sim-Chem Division
P.O. Box 49
Stockport, SK3 ORZ
UNITED KINGDOM

Japan Synthetic Rubber Co., Ltd.
No. 1, 1-Chome, Kyobashi
Chuo-ku, Tokyo 104
JAPAN

The M. W. Kellogg Co.
1300 Three Greenway Plaza East
Houston, TX 77046

Kureha Chemical Industry Co., Ltd.
8-1 Chome, Nihonbashi Horidome-cho
Chuo-ku Tokyo 103
JAPAN

Kyowa Hakko USA Inc.
521 Fifth Avenue
New York, NY 10017

Lankro Chemicals Ltd.
Bentcliffe Works
Salters Lane
Eccles, Manchester M300 BH
ENGLAND

The Leonard Process Co., Inc.
37 W. 37th Street
New York, NY 10018

Licensintorg (USSR)
Licensor:
Davy Powergas Ltd.
8 Baker Street
London, W1M 1DA
ENGLAND

Linde AG
Carl von Linde Strasse 6
8023 Hoellriegelskreuth
FEDERAL REPUBLIC OF
GERMANY

Lonza/First Chemical Corp.
Licensor:
First Mississippi Corp.
700 North Street
Jackson, MS 39205

The Lummus Company
1515 Broad Street
Bloomfield, NJ 07003

American Lurgi Corp.
5 East 42nd Street
New York, NY 10017

Marathon Oil Co.
539 South Main Street
Findlay, OH 45840

Joseph Meissner GmbH
Licensor:
C&I/Girdler Inc.
P.O. Box 174
1721 South Seventh Street
Louisville, KY 40201

Mitsubishi Gas Chemical Co.
The Badger Company, Inc.
One Broadway
Cambridge, MA 02142

Mitsui Toatsu Chemicals, Inc.
Kasumigaseki Bldg.,
3-2-5 Kasumigaseki, Chiyoda-ku
Tokyo 100
JAPAN

MoDoKemi AB
S-444 01 Stenungsund 1
SWEDEN

Monochem Inc.
P.O. Box 488
Geismar, LA 70734

Monsanto Enviro-Chem Systems Inc.
800 North Lindbergh Boulevard
St. Louis, MO 63166

Montedison S.p.A.
Brev/Lic
Lorgo Donegani 1/2
20121 Milano,
ITALY

Montefibre S.p.A.
Via Pola 14
20124 Milano
ITALY

Naphthachemie
203 rue du Faubourg
Saint-Honore
Paris 8e
FRANCE

Nihon Yuki Co., Ltd.
Gamo 3.030
Koshigaya, Saitama,
JAPAN

Nipak, Inc.
301 South Hardwood St.
Box 2820
Dallas, TX 75221

Nippon Shokubai Kagaku
Kogyo Co., Ltd.
(English name)
Japan Catalytic
Chemical Industry Co.
Licensor:
Simon-Carves Ltd.
Sim-Chem. Division
P.O. Box 49
Stockport, SK3 ORZ
UNITED KINGDOM

Nippon Soda Co.
Shin-Ohtemachi Bldg.,
2-2-1 Ohtemachi, Chiyoda-ku
Tokyo 100
JAPAN

Nippon Steel Chemical Co.
6-17-2 Ginza, Chuo-ku
Tokyo 104
JAPAN

Nippon Zeon Co., Ltd.
Furukawa Sogo Bldg.
6-1 Marunouchi 2-Chome
Chiyoda-ku, Tokyo
JAPAN

Nissan Chemical Industries, Ltd.
Kowa Hitotsubashi Bldg.
3-7-1 Kanda-Nishiki-cho,
Chiyoda-ku, Tokyo 101
JAPAN

Petro-Tex Corp.
8600 Park Place
Houston, TX 77017

Pfaudler Co., The Sybron Corp
1000 West Avenue
Rochester, NY 14603

Phillips Petroleum Co.
Bartlesville, OK 74003

PPG Industries, Inc.
Chemical Division
One Gateway Center
Pittsburgh, PA 15222

Polysar Ltd.
Sarnia, Ontario
CANADA

Products Azole
Licensor:
Foster Wheeler Corp.
110 South Orange Avenue
Livingston, NJ 07039

Reichold Chemicals, Inc.
RCI Building
White Plains, NY 10602

Rhone Progil
Service Licenses de Procedes
25, Quai Paul-Doumer
F 92408 Courbevoie
FRANCE

Ruhrchemie AG
Licensor:
Hoechst-Uhde Corp.
550 Sylvan Avenue
Englewood Cliffs, NJ 07632

Ruhrchemie/FWH Farbwerke
Licensor:
Hoechst Uhde Corp.
550 Sylvan Avenue
Englewood Cliffs, NJ 07632

Scientific Design Co., Inc.
Two Park Avenue
New York, NY 10016

Selas Corp. of America
Dresher, PA 19025

Shell Development Co.
One Shell Plaza
P.O. Box 2463
Houston, TX 77001

Sherwin Williams Chemicals
P.O. Box 6520
Cleveland, OH 44101

Shikoku Kasei Co., Ltd.
147 Minato-machi Marugame-City
Kagawa-Pre.,
JAPAN

Simon-Carves Ltd.
Sin-Chem Division
P.O. Box 49
Stockport, SK 3 ORZ
UNITED KINGDOM

SNAM Progetti S.p.A.
20097 San Donato Milanese
ITALY

SNIA Viscosa S.p.A.
via Montbello
18-20121 Milano
ITALY

SNPA
Societe Nationale des Petroles
D'Aquitaine
Tour Aquitaine
Cedex No. 4
Paris-La Defense 92080
FRANCE

Solvay & Cie
Administration Centrale
33, Rue du Prince Albert
Brussels
BELGIUM

Stamicarbon bv
P.O. Box 10
Geleen
THE NETHERLANDS

The Standard Oil Co.
Midland Building
Cleveland, OH 44115

Standard Oil Co. (Indiana)
910 South Michigan Avenue
Chicago, IL 60605

Stauffer Chemical Co.
Westport, CT 06880

Stone & Webster Corp.
One Penn Plaza
250 West 34th Street
New York, NY 10001

Sumitomo Chemical Co., Ltd.
15 5-Chome Kitahama
Higashi-ku, Osaka 541
JAPAN

Technip
(Compagnie Francaise D'Etud
et de Construction)
232 Ave. Napoleon Bonaparte
92500-Rueil-Malmaison
FRANCE

Texaco Development Corp.
135 East 42nd Street
New York, NY 10017

Tokuyama Soda Co., Ltd.
No. 4-5, 1-Chome
Nishi-Shimbashi, Minato-ku
Tokyo
JAPAN

Toms River Chemical Corp.
P.O. Box 71
Toms River, NJ 07853

Toray Industries, Inc.
2, Nihonbashi-Muromachi
2-Chome
Chuo-ku, Tokyo 103
JAPAN

Total-Compagnie Francaise
de Raffinage
Owners (50-50):
Cie Francaise des Petroles
Cie Francaise de Raffinage
5 Rue Michel-Ange
75 Paris 16e
FRANCE

Toyo Toatsu Industries, Inc.
Licensor:
Mitsui Toatsu Chemicals, Inc.
Kasumigaseki Bldg.
3-2-5 Kasumigaseki, Chiyoda-ku
Tokyo 100
JAPAN

Toyo Soda Manufacturing Co., Ltd.
Toso Bldg., 7-7, 1-Chome,
Akasaka
Minato-ku, Tokyo
JAPAN

Ube Industries Ltd.
Products Development Dept.
7-2, Kasumigaseki,
Chiyoda-ku, Tokyo
JAPAN

UCB, S.A.
Chaussee de Charleroi 4
B-1060 Brussels
BELGIUM

Friedrich Uhde GmbH
Degginstrasse 10-12
4600 Dortmund
FEDERAL REPUBLIC OF
GERMANY

Union Carbide Corp.
Chemicals & Plastics Div.
270 Park Avenue
New York, NY 10017

Union Carbide Corp.
Linde Division
270 Park Avenue
New York, NY 10017

Uniroyal International
1230 Avenue of the Americas
New York, NY 10020

United Chemicals & Coke, Ltd.
Licensor:
Foster Wheeler Corp.
110 South Orange Avenue
Livingston, NJ 07039

U.S. Industrial Chemicals Co.
Div. of National Distillers and
Chemical Corp.
99 Park Avenue
New York, NY 10016

Universal Oil Products Co. (UOP)
10 UOP Plaza
Algonquin & Mt. Prospect Roads
Des Plaines, IL 60016

Veba-Chemie AG
Postfach 45
466 Gelsenkirchen-Buer
FEDERAL REPUBLIC OF
GERMANY

von Heyden/Wacker
Licensor:
Wacker Chemie GmbH
8 Muenchen 22
Prinzregentenstrasse 22 Postfach
FEDERAL REPUBLIC OF
GERMANY

Vulcan-Cincinnati, Inc.
1329 Arlington Street
Cincinnati, OH 45225

Wacker Chemie GmbH
8 Muenchen 22
Prinzregentenstrasse 22 Postfach
FEDERAL REPUBLIC OF
GERMANY

Welsbach Corp.
56 Haddon Avenue
Haddonfield, NJ 08033

C. W. Witten
Now owned by:
Dynamit Nobel AG
Troisdorf
FEDERAL REPUBLIC OF
GERMANY

Zimmer AG
60, Postfach 600 102
6 Frankfurt (M)
FEDERAL REPUBLIC OF
GERMANY

BIBLIOGRAPHY

Allinger, N. L., *et al. Organic Chemistry* (New York: Worth Publishers, Inc., 1971).

Chemical Engineering. "Sources and Production Economics of Chemical Products,"(New York: McGraw Hill, 1974).

95th Congress, 1st Session. (95-30). "Data Relating to H.R. 3199 (Clean Water Act of 1977)" (Washington, D.C.: U.S. Government Printing Office, 1977).

"Consent Decree–Natural Resources Defense Council vs. Train," 8 ERC 2120-8 ERC 2136, U.S. District Court, District of Columbia.

Earhart, J. P., *et al.* "Extraction of Chemical Pollutants from Industrial Wastewaters with Volatile Solvents," Department of Chemical Engineering, University of California, EPA-600/2-76-220 for Robert S. Kerr Environmental Research Laboratory, Office of Research and Development, U.S. Environmental Protection Agency, Berkeley, CA (December 1976).

Federal Register 42 (197): Part IV (October 12, 1977).

Fieser, L. F., and M. Fieser. *Advanced Organic Chemistry* (New York: Reinhold Publishing Corp., 1961).

Fuller, B., *et al.* "Scoring of Organic Air Pollutants," in *Chemistry, Production and Toxicity of Selected Organic Chemicals,* Rev. 1, Appendices I, II, III, IV, The MITRE Corp., Contract No. 68-02-1495 for the U.S. Environmental Protection Agency (September-October 1976).

Goldstein, R. F. *The Petroleum Chemicals Industry,* 2nd ed. (New York: John Wiley & Sons, Inc., 1958).

Groggins, P. H. *Unit Processes in Organic Synthesis,* 4th ed. (New York: McGraw-Hill Book Co., 1952).

Gruse, W. F., and D. R. Stevens. *Chemical Technology of Petroleum,* 3rd ed. (New York: McGraw-Hill Book Co., 1960).

"Hydrocarbon Processing," *1973 Petrochemical Handbook Issue.* 52(11): 89-200 (1973).

"Hydrocarbon Processing," *1977 Petrochemical Handbook Issue* 56 (11): 115-242 (1977).

Hydroscience, Inc. "Emissions Control Options for the Synthetic Organic Chemicals Manufacturing Industry," Progress Report No. 7 for the Period September 1, 1977-September 30, 1977, prepared for the Office of Air

Quality Planning and Standards, U.S. Environmental Protection Agency, Research Triangle Park, NC, under Contract No. 68-02-2577, Knoxville, TN.

Kent, J. A. *Riegel's Handbook of Industrial Chemistry,* 7th ed. (New York: Van Nostrand Reinhold Co., 1974).

Lowenheim, F. A., and M. N. Moran. *Faith, Keyes and Clark Industrial Chemicals,* 4th ed. (New York: John Wiley & Sons, Inc., 1975).

McCurdy, P. P., Ed. *Chem. Week,* Buyers Guide Issue (October 26, 1977).

Meister, R. I., Ed. *Farm Chemicals Handbook, 1978* (Willoughby, OH: Meister Publishing Co., 1978).

Morrison, R. T., and R. N. Boyd. *Organic Chemistry,* 2nd ed. (Boston, MA: Allyn & Bacon, Inc., 1966).

National Aniline Division, Allied Chemical Corp. *Aniline,* Allied Chemical Corp., NY (1964).

"New Horizons," The Lummus Co., New York (1954).

Processes Research, Inc. "Air Pollution from Nitration Processes and Air Pollution from Chlorination Processes," Task Force No. 22 and 23 under Contract No. CPA 70-1, prepared for the Office of Air Programs, U.S. Environmental Protection Agency, Cincinnati, OH (March 31, 1972).

Radian Corp. Radian Chemical Data Base Report Generator, "Radian Corp., Austin, TX (April 15, 1976).

Shreve, R. *Chemical Process Industries,* 3rd ed. (New York: McGraw-Hill Book Co., Inc., 1967).

Sittig, M. "Pollution Control in the Organic Chemical Industry," Noyes Data Corp., Park Ridge, NJ (1974).

Stanford Research Institute. *1977 Directory of Chemical Producers - United States of America,* Menlo Park, CA (1977).

Waddams, A. L. *Chemicals from Petroleum* (Pearl River, NY: The Noyes Press, Inc., 1962).

Windholz, Ed. *The Merck Index,* 9th ed., Merck & Co., Rathway, NJ (1976).

INDEX